SpringerBriefs in Applied Sciences and Technology

PoliMI SpringerBriefs

More information about this subseries at http://www.springer.com/series/11159
http://www.polimi.it

Mina Akhavan

Port Geography and Hinterland Development Dynamics

Insights from Major Port-cities of the Middle East

POLITECNICO
MILANO 1863

Mina Akhavan
DASTU
Politecnico di Milano
Milan, Italy

ISSN 2191-530X   ISSN 2191-5318   (electronic)
SpringerBriefs in Applied Sciences and Technology
ISSN 2282-2577   ISSN 2282-2585   (electronic)
PoliMI SpringerBriefs
ISBN 978-3-030-52577-4   ISBN 978-3-030-52578-1   (eBook)
https://doi.org/10.1007/978-3-030-52578-1

This Springer imprint is published by the registered company Springer Nature Switzerland AG
The registered company address is: Gewerbestrasse 11, 6330 Cham, Switzerland

To my family in Iran and in Italy

Foreword by Carola Hein

The field of port-city studies is vast. Researchers from many disciplines study the multifaceted structure of port cities, their global function, their economic role, distinctive appearance and their storied history. They do this by investigating various scales, examining buildings, urban districts—notably waterfronts—or the entire port-city region. Their temporal perspective differs, ranging from long-term studies to forward-looking scenarios or designs. Understanding how port-related flows shape spaces, institutions and even cultures over time and finding meaningful ways for politicians, policy-makers and designers to shape the complex, fuzzy territories that they generate, requires a careful discussion of language and methods from many disciplines and stakeholders.[1]

An integrated multidisciplinary conceptualization with buy-in from different stakeholders is urgently needed as port-city regions around the world today face a number of complex problems including climate change mitigation, adaptation to digitization and migration. Such a conceptualization needs to carefully balance global and local aspects: each port-city region has particularities of geography, long-term histories, prior investment, or political and social dynamics. New concepts and methods are needed to address the future. Quantitative analysis of port throughput or economic urban growth needs to be combined with qualitative investigation. Spatializing the current challenges through mapping can serve as a gap-finder, aiding identification of all the potential areas and opportunities and conflicts between port and city[2]. Future approaches to planning, governance and policy making in port-city regions can also no longer be limited to the construction of the hard infrastructure necessary for the functioning of the port; they need to include development of the skill sets and technologies needed for the port and port city of the future.

---

[1] Carola Hein, "The Port Cityscape: Spatial and Institutional Approaches to Port City Relationships," *PortusPlus* 8 (2019).

[2] Carola Hein and Yvonne van Mil, Mapping as Gap-Finder: Geddes, Tyrwhitt and the Comparative Spatial Analysis of Port City Regions, Urban Planning (ISSN: 2183–7635) 2020, Volume 5, Issue 2, Pages 1–15, https://doi.org10.17645/up.v5i2.2803.

The field is too large for one individual. We need frameworks that bring together a larger cohort of researchers who can engage with and support each other while connecting directly to diverse local stakeholders. Young researchers are key to develop this growing field. Their insight and enthusiasm is also much needed. Mina Akhavan has taken on the challenge. Following a careful review of the literature, she theorizes the role of cities in line with their centrality. For port cities, this means their role as an actor situated between other production and consumption hubs. She notably identifies four stages in the historical development of port cities: "Port city, Port area, Port region, Port network," acknowledging that we need to explore port-ity hubs. She aims to overcome the current focus on European, American or Australian cities, by focussing on the port city of Dubai. Her insights on the role of Dubai's free trade zone are particularly relevant for to the concept of the "logistics hub port city" and what it signifies for Dubai. Such an exploration is promising as a step to better understanding complex port-city relations in regard to space, institutions and ultimately culture.

Carola Hein
Delft University of Technology
Delft, The Netherlands

Foreword by Harvey Molotch

In researching the development of world ports, particularly Dubai and other parts of the Middle-East, Mina Akhavan has made a contribution to urban studies more generally. As she treats it, places of connection—where materials and individuals come together from distant and varied places—itself has consequence. In this, she follows the classic model. But, here she goes beyond that model, showing the results as including transformation of local infrastructure, civic development and rise in economic power. Events at the port and through the port radiate out as new forms of both physical and social infrastructure. The port is not just a consequence of urbanity but, in continuously changing ways, a force in its evolution.

As the world map now shifts routes of relations and scales, the port thus becomes an evolving basis of world integration and new patterning of differentiation. Her central case, Dubai is the extreme but not exceptional instance of the magnifying role of port—in its way a very old-fashioned sort of geo-institution that persists in significance even as it incorporates more cutting-edge technical and organizational elements, including logistic software and global capital evolution.

With Dubai as central case, Akhavan widens inclusion of types of political regime to show how capitalist set-ups merge with types of authority systems not envisioned in many conventional paradigms of how commerce operates in the world. Across her realms come market societies, authoritarian systems, and democracies, all involved and coordinated through systems of trade and integration. Akhavan uses the geo-point, the port, as vantage from which to inspect and understand the larger dynamics.

Meticulous and careful in its documentation, we learn the details of how global infrastructure can take form and come to matter in the world. Conceptually open and rigorous, Akhavan advances interdisciplinary understandings of economy and society.

Harvey Molotch
Emeritus, Professor of Metropolitan Studies
New York University
New York, USA

Preface

My interest on port-city studies goes back to 2011, when I started working on my master's thesis on new urban waterfronts developed along the ex-port sites of European post-industrial port cities. As historic ports in city centres become abandoned, new maritime infrastructures were then constructed far from the urban agglomeration seeking more space and deeper water. Such massive port infrastructures and their widening hinterland are generating new spatial and geographic complexities at different scale. During my three-year doctorate programme—in Spatial Planning and Urban development at Department of Architecture and Urban Studies (DAStU), Politecnico di Milano (Polimi)—I had the opportunity to study and tackle the complex issues regarding the interaction between the city and the development of maritime ports. In fact, a large body of literature already exists on port geography and port cities from varied disciplines, mainly derived from studies on the Western World and fast-developing countries of the far East—such as China. However, there is a geographic gap in port-city literature: we know little about the emerging ports of the Western Asia—also known as the Middle East. In this region, the fast-developing Arab states of the Persian Gulf are becoming key players in the global sphere; not only in the global supply chain but also as key investors and mediators of global commodity flows—the case of terminal operators; Dubai-based DP World is now among the top-5 main global terminal operators, which currently manages a significant amount of the international maritime trade and port traffic.

Framing a theoretical base from a wide body of literature—linking port-city studies, globalization theories and logistics—the particular interest of this book is to illustrate the features and characteristics of the port-city development dynamics, with a focus on the Arab fast-developing city-states of the Middle Eastern region. Considering the fact that the current literature on port cities is quite biased with mono-disciplinary tendencies, the multidisciplinary lenses and mixed methods applied in this study are also considered as innovative aspects worth underlining.

This book is therefore an outcome of my several years of research work, combined with teaching and mentoring activities. I have presented the findings of my studies on this topic in several international conference and then published in peer-reviewed journals. I hope this book would be of interest for scholars, graduate and post-graduate students in different fields of urban studies—spatial planning, urban and regional economics, geography and transport geography—who are seeking a multidisciplinary debates on port geography and spaces of flows generated through maritime activities at the interface between the port and the city. Moreover, I believe that this book will provide useful insights for policy-makers, urban practitioners and technical staff who are dealing with issues related to port-city development.

Here, I would like to use this space to give my special appreciation to Prof. Ilaria Mariotti, my then Ph.D. supervisor and now a dear friend and colleague, for her mentorship and support throughout these years. I am grateful for her hard-working attitude, kindness and patience, which has always inspired and encouraged me. I should also thank my department—DAStU-Polimi—and all my professors, for giving me the chance to study, do research and teaching activities in an interdisciplinary environment. Last but not least, I am greatly indebted to my family. Though been far, none of this would have been possible without the unconditional love, support, patience, encouragement and understanding of my mother, Touran, my brother and sister; Mehdi and Marjan; and of course, my new family in Italy: my husband, Michele, and my newborn baby Arianna.

Milan, Italy   Mina Akhavan

Contents

Chapter 1
State-of-the-Art Studies

1.1 Port-Cities as Laboratories in Urban and Transport Studies

The argument of this book is centred on port-city development trajectories; the changing port and city relationship throughout time and the importance of port infrastructure in logistics and global commodity flows. Many historic cities, that are of great importance today, were originally born along a water body. For such cities, water has been a fundamental source for transportation and trade, which also characterised the main economic activities of the city; a close association between water transport and urban development is therefore apparent. However, in the recent decades, globalization, the global economy, technological advancements and international division of labour have boosted substantial changes in the geography of transportation and its relationship with urbanization trend.

Port-cities are not only cities that exist on the shoreline, but economic centres, which are maritime-based by nature (Broeze 1989). These cities are thriving on maritime flows and connections to trade dynamics, which allow them to take advantage of their strategic location and commercial connections. A port-city is therefore defined '*as a hub in dense networks of maritime connections through which people, goods, ideas and meanings flowed*' (Driessen 2005: 29–30). Ports have historically been of key importance for local economies, but with the growth of containerization and technological advancements, some studies report that ports are losing their direct connection with their cities (Musso et al. 2000). Other studies have also reported that highly trafficked ports may have a lesser impact on the local city's socio-economic development (Bottasso et al. 2013).

Maritime ports and the cities hosting them, i.e. port-cities, have long fascinated scholars—geographers, economists, architects, urban planners, sociologists etc.—as they become centres of exchange where different social and urban environments meet, at the intersection between land and sea. The two components of port-cities: port as a node within a transport system, and city as the central place of a wider spatial system, are mainly considered interconnected in their location, development, activities and issues (Hoyle 1997). The abundance of literature on port geography and port-city studies can be divided as: (i) waterfront re-developments (Akhavan 2019; Breen and Rigby 1996; Bruttomesso 1993; Hein 2011; Hoyle 1989, 2000; Marshall 2004); port-city development models (Akhavan 2017a, b; Bird 1963; Hoyle 1968, 1988, 1989; Lee et al. 2008); (ii) new port infrastructures, hinterland and regional development (Hoyle and Pinder 1981; Notteboom and Rodrigue 2005; Slack 1999); (iii) economic impact of maritime ports (Bottasso et al. 2013; Danielis and Gregori 2013; Ferrari et al. 2010; Hall 2008); (iv) ports as elements within the global supply chain and the logistics activities (Jacobs and Hall 2007; Robinson 2002; Wang et al. 2007); (v) globalization, port hierarchy and the inter-port network structure (Ducruet et al. 2010; Ducruet and Lee 2006; Jacobs et al. 2010); port governance and global terminal operators (Notteboom and Rodrigue 2012; Panayides and Song 2008).

The varied aforementioned studies tend to explore ports and cities as two separate assets, where several actors have minimum interaction. Despite the existence of rich literature on port-city studies, there is rather a divergent perspective in the port-related and urban-related issues. Indeed, ports are not only embedded within the global logistics, but they occupy urban spaces and contribute to local and regional economic growth. Thus, in order to understand the current and future challenges in port-city development, there is a need for a more integrated multidisciplinary approach. Moreover, the existing literature on this topic is predominantly based on the Western and East Asian context. Although the growing port-cities of the Middle East are investing on modernizing and developing new ports, this part of the world is yet to be studied.

Since the 1960s, due to the major advancements and innovation in transportation technology (e.g. containerization) and economic restructuring, scholars from different discipline show a growing interest in studying the changing role and function of the ports at different scales (Basu 1985; Broeze 1989; Hoyle and Hilling 1984; Lawton and Lee 2002; Notteboom and Rodrigue 2005). More recently, comparative studies on European and Asian world has also focused on understanding the port development with relation to the spatial changes of the city (Ducruet 2006; Ducruet and Jeong 2005). In fact, more than half of the century of change in port infrastructures and shipping industry has deeply restructured the interaction between the city and its port.

The connection between ports and their role in logistics has become the subject of study for scholars, some of which have created the concept of port-centric logistics and port regionalization (Bichou and Gray 2004; Carbone and Martino 2003; Heaver 2002; Mangan et al. 2008; Notteboom and Rodrigue 2005; Panayides and Song 2008; Robinson 2002). On this matter, a study by Robinson (2002) emphasized the restructuring of supply chains and logistics where ports were ingrained, while also underlining the shortage of existing paradigms of the time to manage the relationship between port functions and authorities. Robinson's study introduced the new idea of considering ports themselves as elements in the logistics and value-added chains.

Ports are only one of the various components that require the integration of many sectors in the supply chains in order to develop logistics hubs. Ports have been widely recognized as the connection between land and sea and therefore play an important role in the global supply chain by providing logistics services far beyond the handling of simple cargo (Pettit and Beresford 2009). Many ports have been forced to adapt to changes in order to remain competitive in the widening network of supply chains that have become available with the rise of globalization. Notteboom and Rodrigue (2005) spoke about the regionalization phase in port evolution trends and enlarged the port hinterland by providing a closer connection to inland freight distribution cores and critical market strategies. Therefore, the port clearly extends beyond the more historic geographic boundaries, towards a connection between both wider market and developing port-related distribution centres (such as free trade zones).

Scholars have used different concepts to argue the relation between the varied forms of infrastructural network and spaces of flows (Hall and Hesse 2013). Graham and Marvin (2001) and Swyngedouw (1993) discuss that the provision of modern networked infrastructures enables new configurations within and between spaces, while producing new economic geography. According to Ducruet and Lee (2006), transportation hubs are becoming disconnected from the main global cities at the top hierarchy. For Fleming and Hayuth (1994) cities that function as intermediary places are neither central places nor gateway cities. Gateways and hubs are also discussed by Hesse (2010) as outcomes of the changing spaces of flows.

Moving beyond the traditional hierarchal approach in understanding the system of global cities, Sigler (2013) proposes the notion of relational cities (using three port cities of Dubai, Doha and Panama city as illustrative case studies) formed by the complexity of flows—of goods, capital and people—strategically situated in the global and regional economic system and own essential intermediary infrastructures (ports, free trade zones, offshore banking centres, etc.; therefore, it resembles gateways and entrepots from spatial and functional points of view.

The importance of ports in the global supply chain has a long history that can be traced back to the beginning of containerization, which required the infrastructural development of modern ports. Newly developed ports in the 1990s were used for the transhipment of containers between vessels and were strategically located on major traffic flows. Gateways such as Rotterdam were expanded for transhipment activities (Ng and Yu 2006). In the evolution models of Western port-cities, Bird (1963) and Hoyle (1989) considered the stages in which a port faces major geographical and technical transformations in order to harbour large container ships. But earlier port research focused on the physical and functional aspects of the port as a city, while placing less emphasis on the port as a node in regional and global transportation networks (Olivier and Slack 2006).

While questioning the ability of transhipment hubs to develop logistics activities, Slack and Gouvernal (2016) explored the relationship between container ports and the logistics supply chain. Their research argues that most transhipment ports have failed in acquiring logistics operations, except for a few cases, like Singapore, where transhipment has been a part of large logistics industry. Regardless of the previous research, some studies have discussed the case of Dubai as a transhipment hub, copying the model of Singapore, underlining its relative success in developing into an integrated logistics centre (Akhavan 2017b). As stated by Hoyle (2011), the primarily function of a port, as a transport mode within a multimodal system, is transport integration. However, a port facilitated by relevant infrastructures may perform as an urban centre, to generate employment and added value, redefining the conventional relation at the local, regional and global level.

1.2 A Case-Study Approach

Within the literature, the port-city relationship has been investigated by means of a case-study approach (Frémont and Ducruet 2005; Hoyle 1983, 2011; Lee et al. 2008), or through conceptualizing general models of development for the Western developed countries (Bird 1963; Hoyle 1988; Murphey 1989); for the underdeveloped and developing world (Hilling 1977; Hoyle 1968; Taaffe 1963); and for the case of East Asian context (Lee et al. 2008).

A variety of literature states that Singapore is a successful example of an Eastern global port city (Parsa et al. 2002), as it functions as an integrated logistics hub leader (Fernandes and Rodrigues 2009) that other transhipment ports seek to emulate (Slack and Gouvernal 2016). The geographical position in the middle of key trade routes allowed Singapore to develop into a historical trading centre. A large portion of academic insight into logistics functioning of global hub ports specifically target the Western World and East Asian major cities. While quickly developing Arab oil-based countries are becoming key members of the global supply chain, the academic research does not focus on these countries as logistics centres. Dubai, the main hub port-city in the region, has managed to enter the global economic circuit with its massive investment in infrastructure, and in the 1990s it was the first Middle Eastern city to emphasize a world city connectivity (Shin and Timberlake 2000). In the mid-2000s, it appeared as one of the top global transhipment centres along with Hong Kong and Singapore (Baird 2007). Dubai, following a development strategy similar to that of Singapore, serves as home to international hub ports and ancillary infrastructures. This city-state has peaked the attention of international scholars in the interdisciplinary field of urban studies (Akhavan 2017a; Bagaeen 2007; Elsheshtawy 2008, 2004; Jacobs and Hall 2007; Ramos 2010).

Compared to other cities, presence of a port in a city is accompanied by a specific demand on maritime resources and activities, which makes port-cities more complex urban communities. Nevertheless, as an impact of technological advancement and growing level of globalization and transnational flows, the changes in the maritime sector have major underestimated implication on urban development issues. This is even more crucial for the case of fast-developing countries, since the port-city is of national trading importance. The case of Dubai with its fast-developing modern port infrastructure is an interesting case, which shows how the global flows can create complex spatial configurations (Hesse 2008).

By all accounts, Dubai is considered a success story. Despite doubts on its long-term sustainability, the development model to achieve this built-from-scratch 'instant city' (Bagaeen 2007) not only has drawn regional and worldwide attentions of practitioners and city developers, yet also many professional and researchers. Dubai's historical development, from a fishing and pearling village to a global city, as well as its contemporary success story has been well documented in the recent literature (Acuto 2010, 2014; Broeze 1999; Davidson 2008; Elsheshtawy 2010; Hvidt 2009; Ramos 2010). Jacobs and Hall (2007) have studied the factors contributing to the successful emergence of Dubai's port into the global supply chain, to become what they call the 'Singapore of the Middle East'. Sigler (2013) made an interesting study on Dubai's urban change in the global context within the framework of 'relational cities', as a new form of gateway cities and entrepôts. And only few studies have studied Dubai's transport infrastructure and logistics development (Akhavan 2017a, b; Fernandes and Rodrigues 2009; Keshavarzian 2010; Thorpe and Mitra 2011; Ziadah 2018).

"In Dubai everything is bigger". We often hear this statement with reference to single architectural and urban projects—namely, the world's tallest building (Burj Khalifa); the world's largest mall (Dubai Mall); and the world's largest indoor theme park—as well as to infrastructures. The latter include the largest man-made maritime port and two major airports, which are regionally the most important and among the world's top ten. It is worth mentioning that Dubai's port hinterland homes the world's largest logistics corridor, which links sea, land and air, bridging the Jebel Ali Port—a major hub-port; Jebel Ali Free Zone—host to more than 6500 companies; and Dubai World Central—home to Al Maktoum International airport, which upon completion, will be the world's largest airport.

To represent the influence of Dubai on Middle Eastern urbanism, Elsheshtawy (2010) introduced the term "Dubaization", alluding to the proliferation of megaproject development strategy adopted by Dubai real estate investors in other cities, both regionally—namely Cairo—and globally—such as the Panama City. Consequently, "laboratories like Dubai have revealed how agents successfully combined capital, labour, land, and a favoured position on the global scale to develop aggressively both site and situation" (Hesse 2010: 87). Other authors refer to these urbanization and regionalization processes as the 'Dubai model' (Hvidt 2009). One can apply, by similitude, the same concepts to explore port-city relationships. However, we should avoid any single 'Gulf City Model' as also depicted by Molotch and Ponzini (2019).

As a famous touristic and business destination, Dubai is generally recognised and cited for its high-rise buildings, mega shopping malls and luxury hotels; historically known as the 'City of Merchants', Dubai today has less fame as a port-city, and therefore little is known regarding the interaction between the port and the city-state. In this book I will discuss that Dubai's maritime ports and global trade has been central to its development from historical times, and even today the port infrastructure plays a key role in positioning this city within the global network of flows. By conducting a case-study approach, Dubai's development is analysed as a unique case, not least in the Middle East. Though the starting engine for Dubai's modern infrastructural construction is due to the oil revenues, yet limited natural resources led to an economic diversification, followed by hub-making strategies (such as transshipment hub port, trading hub, logistics hub, tourism hub, and etc.). This is well demonstrated by the government-led strategies, through a series of plans and megaproject constructions. This stands as further evidence for the multi-scalar 'transnational Dubai', and the link between global flows and local/regional implications. Thus, the case is critical to study, not only because it may provide insights to be generalized in other post-oil cities of the region, but also for its unique model in developing from an a regional entrepôt into a major global hub port-city.

This book therefore seeks a two-folded objective. On the one hand, it makes an attempt to contribute to the existing literature and to fill the literature gap on the Middle Eastern studies and to broaden our understanding beyond the Western based theories on the contemporary port-city development issues. On the other hand, this study seeks to illustrate the features and characteristics of the origins, current dynamics and future of port-cities at the basis of Dubai's development pattern. From an urban planning perspective, it is expected that the research findings will provide a comprehensive basis for the actors dealing with the port-city development issues, in the case of major transshipment hub ports. Moreover, it is intended that this study will lead to future research avenues for an integrated planning in both logistics and local/regional development, with the aim to increase the local benefits.

1.3   Structure of the Book

The six chapters of this book is about port-cities and more specifically the spatial and socio-economic interaction between the port-activities and city growth in a particular context. The review of the vast literature is critically discussed in Chaps. 2 and 3. The conceptualized model of the Western and Eastern port-city evolution is introduced in Chap. 2 in order to understand the factors and evidence in weakening and/or strengthening the port-city relationship. This chapter concludes by clarifying the scientific relevance of this topic of study and positioning this book within the existing literature. Chapter 3 explores and discusses varied literature centred on the impact of globalization on restructuring the role of ports. The concept of gateway-cities, entrepôts and hub-cities are used to study the role of other cities within the system of global connections. This chapter concludes by underlining the need for a more integrated approach in port-city studies, in order to understand the current and future challenges in port-cities.

Given the strategic location and privileged by the natural wealth, since the 1970s, among the Arabic city-states, Dubai has developed with an unprecedented pace to emerge as a regional hub, while mediating the flows of people, goods, capital, knowledge and ideas. Due to the short history of this fast development, and the emergence of the post-oil urbanism in the Persian Gulf Region, questions arise concerning the origins, factors and characteristics of such pattern of growth and its future. Empirical findings are presented in Chaps. 4 and 5. To study this transformation, a 4-phased port-city development model is introduced and discussed in Chap. 4. Here the aim is to provide an understanding of historical and contemporary port-city relationship in the case of Dubai, with the maritime port has evolving from an entrepôt to a transshipment hub and therefore an important logistics centre within the global supply chain system. Chapter 5 follows the discussion on the previous chapter and analyses the government-led strategies to make the modern Dubai; though the development is largely attributed to oil revenues, that were subsequently invested in the construction of infrastructures, the growing awareness of oil availability, the development of megaprojects and free zones are among the strategies towards diversifying the economy since the 1980s. Here, the effectiveness of these strategies is analysed through a series of data drawn mainly from official statistics centre. Here, special attention is reserved to trade-oriented infrastructure and more especially free zones that started developing around the ports and currently spreading all over the city.

The concluding Chap. 6 provides an overall overview of the arguments. Once again, the twofold aim of the book is underlines and attention is drawn to the importance of studying the emerging hub-port and logistics centres in the Middle Eastern Region that are playing key role in the global trade. Some policy implications and future research lines brings the book to an end.

References

Acuto M (2010) High-rise Dubai urban entrepreneurialism and the technology of symbolic power. Cities 27(4):272–284. https://doi.org/10.1016/j.cities.2010.01.003

Acuto M (2014) Dubai in the 'middle.' Int J Urban Reg Res 38(5):1732–1748. https://doi.org/10.1111/1468-2427.12190

Akhavan M (2017a) Development dynamics of port-cities interface in the Arab middle eastern world—the case of Dubai global hub port-city. Cities 60(part A):343–352 https://doi.org/10.1016/j.cities.2016.10.009

Akhavan M (2017b) Evolution of hub port-cities into global logistics centres. Int J Transp Econ 44(1):25–47. https://doi.org/10.19272/201706701002

Bagaeen S (2007) Brand Dubai: the instant city; or the instantly recognizable city. Int Plan Stud 12(2):173–197. https://doi.org/10.1080/13563470701486372

Baird A (2007) The development of global container transhipment terminals. In: Wang J, Olivier D, Notteboom T, Slack B (eds) Ports, cities and global supply chains. Ashgate, pp 69–87

Basu DK (1985) The rise and growth of the colonial port cities in Asia. University Press of America

Bichou K, Gray R (2004) A logistics and supply chain management approach to port performance measurement. Marit Policy Manag 31(19):47–67. https://doi.org/10.1080/0308883032000174454

Bird J (1963) The major seaports of the United Kingdom. Hutchinson, London

Bottasso A, Conti M, Ferrari C, Merk O, Tei A (2013) The impact of port throughput on local employment: evidence from a panel of European regions. Transp Policy 27:32–38. https://doi.org/10.1016/j.tranpol.2012.12.001

Breen A, Rigby D (1996) The new waterfront: a worldwide urban success story. McGraw-Hill, London

Broeze F (1989) Brides of the sea: port cities of Asia from the 16th–20th centuries. University of Hawaii Press, Honolulu

Broeze F (1999) Dubai: from creek to global port city. In: Fischer LR, Jarvis A (eds) Harbours and havens: essays in port history in honour of Gordon Jackson. International Maritime Economic History Association

Bruttomesso R (1993) Waterfronts: a new frontier for cities on water. International Centre Cities on Water, Venice

Carbone V, Martino MD (2003) The changing role of ports in supply-chain management: an empirical analysis. Marit Policy Manag 30:305–320. https://doi.org/10.1080/030888303200014 5618

Danielis R, Gregori T (2013) An input-output-based methodology to estimate the economic role of a port: the case of the port system of the Friuli Venezia Giulia Region, Italy. Marit Econ Logist 15(2):222–255

Davidson CM (2008) Dubai: the vulnerability of success. Columbia University Press, New York

Driessen H (2005) Mediterranean port cities: cosmopolitanism reconsidered. Hist Anthropol 16(1):129–141. https://doi.org/10.1080/0275720042000316669

Ducruet C (2006) Port-city relationships in Europe and Asia. J Int Logist Trade 4(2):13–36

Ducruet C, Jeong O (2005) European port-city interface and its Asian application. Research report 17. Korea Research Institute for Human, Anyang

Ducruet C, Lee S (2006) Frontline soldiers of globalisation: port–city evolution and regional competition. GeoJournal 67(2):107–122

Ducruet C, Lee S-W, Ng AKY (2010) Centrality and vulnerability in liner shipping networks: revisiting the Northeast Asian port hierarchy. Marit Policy Manag 37(1):17–36. https://doi.org/10.1080/03088830903461175

Elsheshtawy Y (2004) Redrawing boundaries: Dubai, an emerging global city. In: Elsheshtawy Y (ed) Planning Middle Eastern cities: an urban kaleidoscope in a globalizing world. Routledge, London, pp 169–199

Elsheshtawy Y (2008) The evolving Arab city: tradition, modernity and urban development. Routledge, London

Elsheshtawy Y (2010) Dubai: behind an urban spectacle. Routledge, Abingdon

Fernandes C, Rodrigues G (2009) Dubai's potential as an integrated logistics hub. J Appl Bus Res 25(3):77–92

Ferrari C, Percoco M, Tedeschi A (2010) Ports and local development: evidence from Italy. Int J Transp Econ 37(1):1–26

Fleming DK, Hayuth Y (1994) Spatial characteristics of transportation hubs: centrality and intermediacy. J Transp Geogr 2(1):3–18. https://doi.org/10.1016/0966-6923(94)90030-2

Frémont A, Ducruet C (2005) The emergence of a mega-port from the global to the local, the case of Busan. Tijdschr Econ Soc Geogr 4:421–432

Graham S, Marvin S (2001) Splintering urbanism: networked infrastructures, technological mobilities and the urban condition. Routledge, London

Hall PV (2008) Container ports, local benefits and transportation worker earnings. GeoJournal 74(1):67–83. https://doi.org/10.1007/s10708-008-9215-z

Hall PV, Hesse M (2013) Cities, regions and flows. Routledge studies in human geography. Routledge, Oxford

Heaver TD (2002) The evolving roles of shipping lines in international logistics. Int J Marit Econ 4(2002):210–230. https://doi.org/10.1057/palgrave.ijme.9100042

Hein C (2011) Port Cities: dynamic landscapes and global networks. Routledge, London

Hesse M (2008) The city as a terminal. In: Transport and mobility series. Ashgate, p 207

Hesse M (2010) Cities, material flows and the geography of spatial interaction: urban places in the system of chains. Glob Netw 10(2010):75–91. https://doi.org/10.1111/j.1471-0374.2010.002 75.x

Hilling D (1977) The evolution of a port system—the case of Ghana. Geography 62(2):97–105

Hoyle BS (1968) East African seaports: an application of the concept of "anyport." Trans Inst Br Geogr 44:163–183

Hoyle BS (1983) Seaports and development: the experience of Kenya and Tanzania. Gordon and Breach

Hoyle BS (1988) Development dynamics at the port-city interface. In: Hoyle BS, Pinder D, Husain S (eds) Revitalising the waterfront: international dimensions of dockland redevelopment. Belhaven Press, Great Britain, pp 3–19

Hoyle BS (1989) The port—city interface: trends, problems and examples. Geoforum 20(4):429–435. https://doi.org/10.1016/0016-7185(89)90026-2

Hoyle BS (1997) Ports, port cities, and coastal zones: development, interdependence and competition in East Africa. Academie royale des sciences d'outre-mer

Hoyle BS (2000) Global and local change on the port-city waterfront. Geogr Rev 90(3):395–417. https://doi.org/10.1111/j.1931-0846.2000.tb00344.x

Hoyle BS (2011) Seaports and development: the experience of Kenya and Tanzania. Routledge

Hoyle BS, Hilling D (1984) Seaport systems and spatial change: technology, industry, and development strategies. Wiley, p 481

Hoyle BS, Pinder D (1981) Cityport industrialization and regional development: spatial analysis and planning strategies. Pergamon Press

Hvidt M (2009) The Dubai model: an outline of key development-process elements in Dubai. Int J Middle East Stud 41:418a. https://doi.org/10.1017/S0020743809091508

Jacobs W, Hall PV (2007) What conditions supply chain strategies of ports? The case of Dubai. GeoJournal 68(4):327–342. https://doi.org/10.1007/s10708-007-9092-x

Jacobs W, Ducruet C, De Langen P (2010) Integrating world cities into production networks: the case of port cities. Glob Netw 10(1):92–113. https://doi.org/10.1111/j.1471-0374.2010.00276.x

Keshavarzian A (2010) Geopolitics and the genealogy of free trade zones in the Persian Gulf. Geopolitics 15(December 2014):263–289. https://doi.org/10.1080/14650040903486926

Lawton R, Lee WR (2002) Population and society in western European port cities, C.1650–1939. Liverpool University Press

Lee S-W, Song D-W, Ducruet C (2008) A tale of Asia's world ports: the spatial evolution in global hub port cities. Geoforum 39(1):372–385. https://doi.org/10.1016/j.geoforum.2007.07.010

Mangan J, Lalwani C, Fynes B (2008) Port-centric logistics. Int J Logist Manag 19:29–41. https://doi.org/10.1108/09574090810872587

Marshall R (2004) Waterfronts in post-industrial cities. Taylor & Francis

Molotch H, Ponzini D (eds) (2019) The new Arab urban: gulf cities of wealth, ambition, and distress. New York University Press, New York

Murphey M (1989) On the evolution of the port city. In: Broeze F (ed) Brides of the sea: port cities of Asia from the 16th–20th centuries. University of Hawaii Press, Honolulu, pp 223–245

Musso E, Benacchio M, Ferrari C (2000) Ports and employment in port cities. Int J Marit Econ 2(4):283–311

Notteboom T, Rodrigue J (2005) Port regionalization: towards a new phase in port development. Marit Policy Manag 32(3):297–313

Notteboom T, Rodrigue J-P (2012) The corporate geography of global container terminal operators. Marit Policy Manag 39(3):249–279. https://doi.org/10.1080/03088839.2012.671970

Olivier D, Slack B (2006) Rethinking the port. Environ Plan A 38(8):1409–1427. https://doi.org/10.1068/a37421

Panayides PM, Song D-W (2008) Evaluating the integration of seaport container terminals in supply chains. Int J Phys Distrib Logist Manag 38:562–584. https://doi.org/10.1108/0960000830810900969

Parsa A, Keivani R, Sim LL, Ong SE, Younis B (2002) Emerging global cities: comparison of Singapore and the cities of United Arab Emirates. Real Estate Issues Am Soc Real Estate Counsel 27:95–101

Pettit SJ, Beresford AKC (2009) Port development: from gateways to logistics hubs. Marit Policy Manag 36(3):253–267. https://doi.org/10.1080/03088830902861144

Ramos SJ (2010) Dubai amplified: the engineering of a port geography. Ashgate

Robinson R (2002) Ports as elements in value-driven chain systems: the new paradigm. Marit Policy Manag 29(3):241–255. https://doi.org/10.1080/03088830210132623

Shin K-H, Timberlake M (2000) World cities in Asia: cliques, centrality and connectedness. Urban Stud 37(12):2257–2285. https://doi.org/10.1080/00420980020002805

Sigler TJ (2013) Relational cities: Doha, Panama City, and Dubai as 21st century entrepôts. Urban Geogr 34(5):612–633. https://doi.org/10.1080/02723638.2013.778572

Slack B (1999) Satellite terminals: a local solution to hub congestion? J Transp Geogr 7(4):241–246. https://doi.org/10.1016/S0966-6923(99)00016-2

Slack B, Gouvernal E (2016) Container transshipment and logistics in the context of urban economic development. Growth Change 47(3):406–415. https://doi.org/10.1111/grow.12132

Swyngedouw E (1993) Communication, mobility and the struggle for power over space. In: Giannopoulos GA, Gillespie AE (eds) Transport and communications innovation in Europe. Belhaven Press, pp 305–325

Taaffe EJ (1963) Transport expansion in underdeveloped countries: a comparative analysis. Geogr Rev 53(4):503–529

Thorpe M, Mitra S (2011) The evolution of the transport and logistics sector in Dubai. Glob Bus & Econ Anthol II(2):342–353

Wang JJ, Olivier D, Notteboom T, Slack B (2007) Ports, cities, and global supply chains. Ashgate, Aldershot

Ziadah R (2018) Transport infrastructure and logistics in the making of Dubai inc. Int J Urban Reg Res 42(2):182–197. https://doi.org/10.1111/1468-2427.12570

Chapter 2
Changing Interaction Between the Port and the City. West Versus East

2.1 Port-City Development Trajectories: Western Developed Countries Versus the Eastern World

The complex relationship and interaction between the maritime port and the city has been studied and well documented in the literature (Fujita and Mori 1996; Gleave 1997; Hoyle 1983; Hoyle and Hilling 1970; Omiunu 1989). Some early studies have used simple diagrams or models to analyse and interpret the evolution of ports and the regional system in which they combine (Bird 1963), while others introduced models to describe the changing and mainly weakening, connection between the port and the city in the Western Developed World (Hoyle 1989). In contrast, some studies on Eastern cases (such as Singapore and Hong Kong) show a growing consolidation between the two entities (Lee et al. 2008). The latter model is also confirmed in other major global hub-port cities of the Middle East, such as Dubai (see Chap. 4).

The pioneering work by James Bird in the early 60s, which is still referenced today, shed new lights on the studies on port infrastructure development and its changing relation with the city. Based on his studies on British ports (Bird 1963) introduces the *Anyport* model and discuses that a port may experience three main development phases of: (1) *setting* (up to the nineteenth century—pre-industrial era): located at the historic urban centre, the port is formed and functions based on its geographical features, with port-related activities (warehousing) adjacent to the port; (ii) *expansion* (since the nineteenth till early twentieth century—industrial revolution): the port hinterland expands; quays and new ducks are constructed to cope with the industrialization, which then restructures port functions and port-related activities; railways and roadways connects ports to the greater region; (iii) *specialization* (since the mid-twentieth century—technological era and globalization): containerization and technological advancement necessitates larger specialized piers to handle larger and container ships. Therefore, the need for more space and handling capacities has forced ports, originally located at the heart of the historic city centre, to relocate mainly towards areas with more space to construct specialized infrastructures and enable deeper dredging. Intermodal and Multimodal transportation further connects

M. Akhavan, *Port Geography and Hinterland Development Dynamics*,
PoliMI SpringerBriefs, https://doi.org/10.1007/978-3-030-52578-1_2

the land, air and the sea. The export site may either become completely abandoned and inactive for port functions (commodity traffic) or still be partially active for limited cargo handlings.

Table 2.1 highlights the key elements behind the *Anyport* model, from the original setting of the port up to the mid-twentieth century. Bird (1971) discussed that *Anyport*

Table 2.1 Bird's Anyport model stages of port development

Main development phases	Key characteristics	Period	Illustration
Setting Original setting of the port is based on its geographical features; basic terminal facilities; adjacent to the port, the port-related functions were simply warehousing and wholesaling	*Original port*; mainly a fishing port with trading and shipbuilding activities, composed of several quays	Up to the nineteenth century (peri-industrial era)	
Expansion Industrial revolution led to major changes; port-related activities included industrial activities	*Quays expanded*; to handle the increasing freight and passenger flows *Docks construction*; as a consequence of enlarging ships *Railways and Roadways* ensures connection to the expanded hinterland	Nineteenth–early twentieth century (industrial revolution)	
Specialization Technological advancement and larger ships; modernized and specialized port terminals away from the original site	*Containerization* and *innovation* in the *maritime industry* necessities the construction of *specialized piers* More efficient *intermodal* and *multimodal* transportation *Redevelopment* of the abandoned original port sites for urban functions	Mid-twentieth century (technological and globalization era)	

Source Constructed by the author

was to provide a comparative base for port development; yet, it does not imply that all ports should follow this pattern. However, apart from some detailed local difference, there have been enough similarities to consider this model as a useful concept for describing the general morphological evolution of ports in time and space.

The Anyport model was introduced to identify and explain the stages of development as a mean to examine the port's evolution within the regional and national system; therefore, it is essentially based on the changing physical characteristic of a port. With the growing specialization and new port-infrastructure developments outside the historic city centre, some scholars have updated the original model to fit other regions and also to the contemporary situation. In this regard, Hoyle (1968) made an attempt to show how the Bird's (British-based) Anyport model, can be modified as a framework to analyse the growth pattern of the East African seaports. Based on his study, the East African system is characterised by the similarities and contrasts between the individual terminals, mainly due to the variety of physical, economic and political factors. Hilling (1977) has also analysed the changing pattern of the port activity in the case of Ghana, compared to the so-called 'ideal-typical' sequence of existing models. He emphasised the importance of Bird's Anyport not as a pattern to fit all the ports, but instead to be used as a basis for empirical and comparative studies of port in different regions. Based on his analysis on the three models of the British-based (Bird 1963), the Australian-based (Bird 1968) and the East African-based (Hoyle 1968) he identified three stages of development for ports in Ghana: Surf port, lighterage, and deep-water port (Hilling 1977).

Since the previous models were proved to be weak in describing the changing relationship between the port and its hosting city, and also the features related to the effects of the maritime, technological and logistics development on the urban economy. To overcome this shortage, Hoyle (1988, 1989) proposed a model emphasizing the changing, and mainly weakening, connection between the two realities: '*Economically and geographically, ports and cities have grown apart*' (Hoyle 1989: 430). In his well-known publication '*The port-City interface: Trends, problems and examples*', which has so far being cited the most (based on a search on Scopus with the keywork 'port-city interface', accessed March 2019), he describes a number of factors responsible for this altering land-maritime interrelationships:

- The need for more space, for modernized ports, and deeper dredging, for bigger ships, induced the migration of several terminals towards peripheral areas.
- The advent of containerization and other innovation in the shipping industry led to reduction in traditional port activities and labour forces.
- Advanced hinterland accessibility has enabled the port-related activities to locate further inland and not necessarily in proximity to the port.
- As port infrastructures relocate and the ex-sites become abandoned, spaces open up in city centres that has led to a complex land-use competition among different parties (private stakeholders, local government and the port authority) for the revitalizing the traditional port-city interface for waterfront redevelopment projects (see Sect. 2.3).

Accordingly, Hoyle's modified version of the Anyport-type model describes the port-city development in six main phases (see Table 2.2), underlining the changing port and city relationship instead of merely describing the spatial changes of the port-infrastructure over time: an evolution of the port-city interface (this concept is then reviewed in length in Sect. 2.2), highlighting the main spatial, economic and land-use features and the changes imposed over time. Here, it is evident that the port has always been an essential element of the city economy in different phases, yet with different degrees and modalities.

The introduced models are based on studies carried out on the Western Developed countries and some limited examples in the global south (East African ports). Hoyle (1993: 333), stated that: *'this process can be illustrated from San Francisco to Sydney, from Southampton to Singapore …. Each case is unique, but the underlying principle remain largely the same'*. Yet his model has failed in anticipating different trend in specific regions. Few studies have, nevertheless, considered the port-cities in developing countries (Gleave 1997) and concluded with rather different outcomes from the Western model.

The process of manufacturing moving out from the Western cities to the Eastern colonial port-cities, has affected the port-city relationship in the East, which could

Table 2.2 Six-phases of the Western port-city development

Stage—main characteristics	Period
I. Primitive City-Port Pre-industrial stage; close spatial and functional association between city and port	Ancient/medieval to nineteenth century
II. Expanding City-Port Industrial and commercial growth; ports develop beyond city borders with linear quays and break-bulk industries	Nineteenth–early twentieth century
III. Modern Industrial City-Port Fordism/economies of scale; industrial growth (esp. oil refining) and introduction of containers/ro-ro require separation and increased spaces	Mid-twentieth century
IV. Retreat from the Waterfront Post-Fordism stage; changes in maritime technology induced growth of separated maritime industrial development areas	1960s–1980s
V. Redevelopment of the Waterfront Flexible accumulation; large-scale modern port consumes large areas of land and water space; urban renewal of the original core	1970s–1990s
VI. Renewal of Port-City Links Globalization and intermodalism transform the role of ports; port-city associations renewed: urban redevelopment enhances port-city integration	1990s–2000+

Source Akhavan (2019: 101)

not be explained by the Western models of port-city development. In contrary to Hoyle's model of separation and the notion of port-city interface as the zone of conflict, Lee et al. (2008) proposed the *Asian Consolidation Model*, based on the unique evolutionary process of the Asian global hub port cities (Singapore and Hong Kong). According to their model, the continuation of port activities close to the urban core is an opportunity rather than a threat, to strengthen the port-city linkage and increase its competitiveness.

Based on their detailed study on Hong Kong and Singapore, Lee et al. (2008) proposed the six-stage *Asian consolidation model* with different evolutionary trend compared to the Western based models. This is mainly due to the continuation of port activities in close proximate to the city core, as they became separated in the Western model since the twentieth century. Table 2.3 highlights the main characteristics of the Asian port-city relationship in time. However, it should be underlined that this model is an outcome of a study on major Asian hub port-cities—Singapore and Hong Kong—thus not all Asian cases may follow the same pattern.

Table 2.3 Six-phases of the Asian port-city development

Stage—main characteristics	Period
I. Fishing Costal Village Relatively limited area; small community of natives with self-sufficient local trade	Ancient/medieval to nineteenth century
II. Colonial City-Port Small harbours become ports with a hierarchal structure along the coastal urban system; port facilities expanding as trade increases; growing concentration of industries and population around the urban core	Nineteenth–early twentieth century
III. Entrepôt City-Port Continuing trade expansion and entrepôt function; modern port development and reclamation from sea; important to connect West through maritime trade	Mid-twentieth century
IV. Free Trade Port-City Special economic zones and mega terminals; export-led government policies to attract industries using port facilities through tax-free procedures and low-cost labour	1960s–1980s
V. Hub-Port City Increasing port productivity due to economies of scale, hub functions and territorial pressure close to the urban core; city as global centre for tertiary and tourism activities; Western-like waterfront projects	1970s–1990s
VI. Global Hub Port-City Rise of integrated global logistical system to handle the rapid containerization; maintained port activity and new port building due to rising costs in the hub, possible hinterland expansion	1990s–2000+

Source Constructed by the author, adapted from Lee et al. (2008)

The port-city development dynamics in the Middle East, has been studied by Akhavan (2017). A four-stage has been used to describe the evolving relationship between the port and the city in the case of Dubai: (i) The fishing village and the advent of a free port (1900s–1950s); (ii) The entrepôt port-city (1960s–1970s); (iii) The transshipment hub port city (1980s–1990s); (iv) The logistics hub port city (2000s–present) (see Chap. 4). This model is characterised, on the one hand, by the massive influx of oil revenues invested on modern transport infrastructures— starting with maritime ports; on the other hand, through developing the free port into varied free zones to allow foreign investments. Despite sharing some common aspects with the above introduced Asian model, the Dubai model demonstrates a particular port-city development.

Port-development models show dissimilar trends in advanced and developing countries, since the port characteristics in these two regions differ (Hoyle 1983). However, one major common aspect between the two regions is the 'new port-infrastructure' extended beyond the historical city. The shift of port activities towards outer areas is due to the need for more space and accessibility, for continuing growth in trading. In Western port-cities, traditional port spaces have been mainly re-developed for more urban functions, yet in the Asian model the former port sites are still crucial for port-related activities and serve the international trade. Therefore, the inner and outer port sites are working as interdependent complex entities.

Some scholars define contemporary port-cities with respect to the growing separation between the urban and port systems/functions. For the case of England, Bird (1963) outlined the creation of efficient terminals like 'outports' in unconstrained sites. Gleave (1997) studied the role of port activities in shaping the spatial structure of Freetown—Sierra Leone. He argued that although port activities have historically been important, other factors are becoming more significant. As the basis of the urban economy started to develop, the structure of society changed, and the city requires additional functions (Gleave 1997). On the other hand, some scholars believe that despite the on-going changes in the maritime transport system, most of the world's significant ports are occupying urban[1] spaces and populous urban agglomerations (Hall and Jacobs 2012). Thus, the weakening port-city relationship, illustrated in the traditional models of port development, has been accompanied by the creation of new relationship at a broader yet still urban scale.

Apart from some exceptions, this new relation can also be explained through the coincidence of logistics and global city regions, as discussed by O'Connor (2010), through which a number of major ports are situated along service ranges of important commercial centres. Some scholars argue that numerous port-cities continue to flourish even after their historical port-infrastructure becomes less important (Fujita and Mori 1996). Nonetheless, as an effect of globalization, foreign actors and firms have entered the local port system, thus restructured the long operating histories and string local economic ties between the city and its port (Slack and Frémont 2005).

---

[1] The term 'urban' here refers to urban agglomerations of a central city and the surrounding built-up areas that are related by commuting patterns and other daily interactions; thus, the definition is functional rather than administrative.

A study by Bryan et al. (2006) on the case of South Wales shows that port-related industries are regional players and therefore contribute to the economic and social development of the local economy.

Based on the concept of centrality[2] and intermediacy,[3] Ducruet and Jeong (2005) introduces a useful synthetic matrix of port-city relationship based on 'urban centrality' and 'port intermediacy'. This matrix can be helpful in understanding and classifying different port-cities base on the one hand the city's capacity to generate activities, and on the other hand the spatial quality of its transportation system. Figure 2.1 shows the hierarchal combination of the two main orientations (centrality and intermediacy). The *cityport* at the middle of the matrix represents the state of balance between the city size and port functions. While the *coastal town* is a city with limited size and port activities, the *port metropolis* stands at the opposite side with the capacity to maintain its central position to expand in size and generate large number of port functions.

Compared to other transport infrastructures, the empirical evidence on the impact of port-infrastructure on economic growth and employment is less rich. Regarding the interaction between the port function and the urban system, the literature belongs to two opposing ideas of declining/weakening and rising/strengthening relationship. Some studies have assessed and found positive impact of port activity on local growth and employment in the Western World (Castro and Millán 1998; Gripaios and Gripaios 1995; Haezendonck et al. 2000; Notteboom and Winkelmans 2001). In the case of Asian port cities, recent empirical studies on Chinese port-cities also illustrate a strong correlation between the port development and urban economic growth (Ducruet 2006; Ducruet and Jeong 2005).

In contrast to the positive viewpoint, a number of researchers support the idea of declining port's economic benefits, while the city burdens more costs. Some scholars argue, in the European cases, the relative relation between the reducing number of port employment with the changing spatial pattern of port service-based economies and restructuring port hinterland, port networks and other means of transport (Hesse and Rodrigue 2004). Though historically ports were important backbone of local employment structure, yet with the rising containerization and technological advancement, the ports worldwide are significantly losing their direct employment effect (Musso et al. 2000). The port system may nevertheless play a relevant microeconomic role

---

[2] *Centrality* is referred to the level of attraction and traffic-generating power, which is related to the city's size and significant intermodal activities. The notion of centrality for port-terminals is also related to the *central place theory,* a concept in Urban Geography, which introduces centrality as an aspect in determining urban hierarchy. Centrality is equated with the size of its terminal; thus, many major terminals arise out of cities with more centrally located markets. An example is the port of Shanghai, which is supported by a large market, industrial and manufacturing base (Rodrigue et al. 2013: 135).

[3] Although *intermediacy* may imply to the geographical meaning of 'in betweenness', however is a spatial quality related to the specific context's transportation systems and terminals for passenger/freight flows. Intermediate place gain advantage of extra traffic when is served by transport carriers and the ability for transshipment (e.g. hub ports) (ibid.).

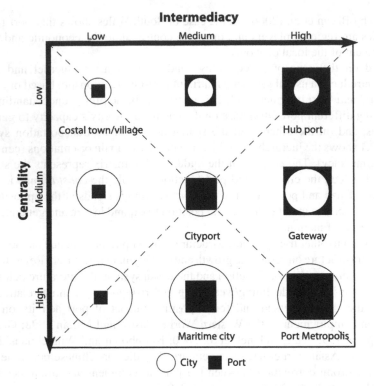

Fig. 2.1 A matrix of port-city relationships. *Source* Redrawn by the author, adapted from Ducruet and Jeong (2005: 9)

within the region, as in the case of Friuli Venezia Giulia Region, studied by Danielis and Gregori (2013) through identifying its industrial characteristics.

2.2 Evolution of the Port-City Interface

The notion of *port-city interface* was introduced by Hayut (1981) as an area in transition, while connecting the port and city through a threefold dimension of changes: spatial, economic and ecological. In spatial terms, the interface zone is the area where port activities and urban functions superimpose, which has been under constant transformation, not only spatially but also functionally (Lee and Ducruet 2009). The port-city interface, from an urban planning point of view, may nevertheless be both problematic, as well as an opportunity for the city, private stakeholders and the port authority. As seen beforehand in the port-city development models, the interface zone was preliminary defined by the geographic proximity of the port activities and the city centre, yet in the contemporary era the changes in this zone are characterised by the migration of port activities away from the traditional site towards more land and deeper water.

For Hoyle (1988, 1989), the port-city interface is an interactive economic system; an area of conflict and/or cooperation and competition, acting as a highly sensitive area, which indeed requires careful planning and assessment concerns prior to development and intervention. He further uses this concept in order to explain the elements involved in the contemporary port-city development, as illustrated in Fig. 2.2:

1. at the interface zone, urban land uses are separated from the maritime functions;
2. port development outside the traditional port-city interface for deep water and more land;
3. port-related industries also relocate to other urban zones, since they are no longer dependent to be located in close proximity to the break-bulk functions;
4. increasing land-use and water-use competition for redevelopment of the waterfront both for urban uses (such as housing and recreational activities) and also maritime based functions (such as water-based facilities);

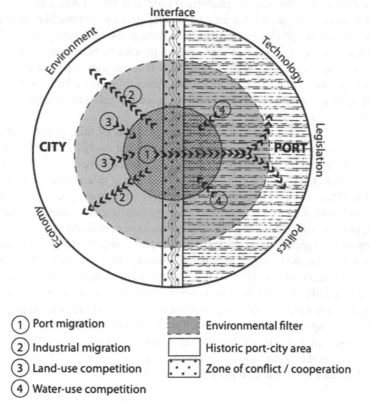

Fig. 2.2 Factors and processes involved in port-city development. *Source* Redrawn by the author, adapted from Hoyle (1989)

5. environmental controls are set at the waterfront zone as a filter to control the redevelopment projects;
6. other factors affecting the overall development of the port-city is related to the technological change, economic and political and institutional situation at different scale.

Hayuth (2007) divides the evolution of port-city interface of the contemporary era, into three main phases: (i) *containerisation*, which had a significant impact on changing the spatial interaction between the port and the city since the 1970s; (ii) *intermodality*, as organizational revolution that allowed the expansion of the hinterland and therefore altered the level of connectivity since the mid-1980s; (iii) *globalization*, which gave rise to the volume of trade and container traffic, also the advent of global port operators (see Chap. 5).

The changing port-city interface in the field of economic geography has been studied by Hall and Jacobs (2012), which argue the 'dynamic' interrelationship between the port and its hosting city. What is interesting is the question pursued by the scholars on *why ports continue to occupy urban spaces?* Using the theories developed by new economic geographers (Fujita and Mori 1996; Krugman 1997), they summarize three possible scenarios for the port-city interaction, in time and space, as the following: (i) cities benefit from their ports; (ii) ports gradually benefit from urban spaces; (iii) ports and urban growth tend to benefit one another. The idea of cities and ports developing separately is then seen more probable than being dependant on one another; yet, at the earlier phases of growth, both entities benefit and only after a certain period, the positive reciprocal relations become exhausted due to all kinds of externalities.

Comparing the European and Asian cases in terms of their port-city interface, here it is worthwhile making reference to the empirical study by Ducruet and Jeong (2005). In Asian cases, due to the importance of coastal market, the sea becomes more significant than in the European port-cities where inland market is dominant. One the one hand, the European territory is experiencing an integration of economies at different levels, thus moving towards one single market and hinterland, defined by multiple ports. On the other hand, the Asian ports have a particular role in the national economy and competitiveness. Moreover, the spatial organization of the European ports, as gateways, are serving a remote hinterland; while in the case of Asian port-cities, as multifunctional nodes, the close hinterland contributes to the concentration of a wide range of port functions. Accordingly, both regions are facing the following challenges: a growing coastal concentration in Asian port-cities is creating issues such as 'lack of spaces'; and Europe is facing a 'growing specialization', followed by inland concentration and rising importance of intermodal transport systems. Despite the difference in the spatial and economic structure, the two regions are developing similar strategies and policies regarding the interfaces zone.

In order to look beyond spatially focused models in confronting the port-city evolution, scholars have considered more complex economic and institutional condition of the current era within a broader setting. An interesting study on Asian cases—with a focus on Chinese ports—has interpreted the evolution of ports in the port-reform era as the port-city interplay with four types of connection: (i) economic and functional relationships; (ii) interactive geographical relationships; (iii) the external network extends by the port to the city; (iv) port-city governance and administrative steering (Wang 2014: 23–24).

Others have made attempts to apply different approaches: a recent study by Hesse (2018) used a relational approach in reading the port-city interface in Europe. He introduces a threefold framework to discuss the wider urban and economic system that links ports and cities: (i) the role of ports as economic accelerators in a wider region; (ii) tourism and cruise industry as a market sector that re-connects port and city; (iii) the importance of semi-public entities as intermediaries and knowledge brokers. Applying a relational approach seems relevant to the study of port-city interface as it allows an understanding of the constitution of relational practices, actors, networks and mobilities as part of a larger territorial scale and complex institutional structure (Storper 1997). Van den Berghe et al. (2018) have also applied this approach to explore the port-city interface in the Netherlands and Belgium: they discuss a form of structural coupling of a specific sector in both cases studies as a higher-level emerging outcome and used this relational concept to describe the interface zone as an interactive complex system, one that is characterized by self-organization, non-linearity and adaptation, while underlining the way ports and transport networks are part of larger urbanized ecosystems which drive change (p. 61).

2.3 Transformation of Ex-Port Sites and New Waterfront Developments

A large body of the literature on port-city relationship and the interface zone is concerned about the contemporary transformation process at the urban waterfront, as also discussed beforehand, which led to a rising competition for mixed-used redevelopment schemes and revitalization of the export site. Although the focus of this book is not on the traditional view of the port-city interface, which has undergone a process of radical spatial transformation as new urban waterfront, yet it is worthwhile providing a quick review of this subject matter as it covers a significant part of the port-city studies.

As discussed in the port-city development models, mainly since the 1960s and 70s, and with the historic port being gradually underutilized, the traditional interface zone became an opportunity for city developers for urban renewal; Baltimore and Boston known as the starting point (Bruttomesso 1993). As the export site revitalization phenomenon becomes increasingly widespread, it is a matter of concern for interdisciplinary studies: geography has played, and continues to play, a leading part

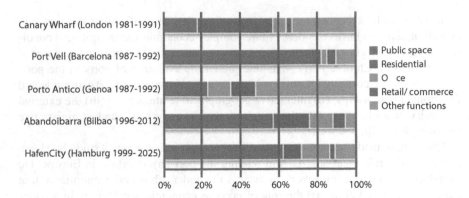

Fig. 2.3 Land-use share of some major European waterfront regeneration projects. *Source* Adapted from Merk (2010: 105)

in these debates—a good measure of interdisciplinary cooperation is long apparent (Hoyle 2000: 402); for architects, urban planners and designers waterfront regeneration has been a major challenge and opportunity for post-industrial port-cities (Marshall 2004), as it becomes a real planning concern (Hein 2011). The export sites for waterfront regeneration projects are therefore considered to be successful and effective as a tool in urban planning and governance to (Breen and Rigby 1996).

Reviewing the varied new urban waterfronts worldwide, one can categories them in two main categories: (i) urban facilities: public space and mixed-used developments (housing, offices, recreational, commercial activities, etc.); (ii) tourism and maritime-based facilities along with a rising demand for berth space and water-related activities such as cruising. Other scholars have also classified waterfront regeneration project based on their development orientation and main functions (Merk 2010). Figure 2.3 illustrates and compares the land use share of some well-known European examples: Canary Wharf (London 1981–1991), Port Vell (Barcelona 1987–1992), Porto Antico (Genoa 1987–1992), Abandoibarra (Bilbao 1996–2012) and HafenCity (Hamburg 1999–expected 2025).

In the new urban waterfronts of Barcelona and Genoa, known to be successful cases, which were regenerated through a systematic use of mega-events (for example, the 1992 Summer Olympics in Barcelona and the 500th anniversary of Columbus's discovery of America—1992 in Genoa), the development was focused on public space creation (about 80%), in the case of Genoa, less than 20% is dedicated to public spaces and instead cultural heritage becomes a major motivation, along with new service and leisure time facilities (Bonfantini 2015; Akhavan 2019).

Many megaproject redevelopments along the waterfronts are coupled with giving back the water to the city, making a new city image, generating economic investments, while attracting city users and tourists to the rather abandoned historic centre. Nevertheless, it is believed that only few projects are truly beneficial socially and economically (Daamen and Louw 2016). According to Hoyle (2000: 415), "the relative success of [waterfront redevelopment] will depend essentially on three things:

integration, integration, and integration. First, integration of past and present; second, integration of contrasting aims and objectives; and third, integration of communities and localities involved." While evaluating a waterfront redevelopment scheme, one question can be raised: *who gains from this transformation, the city or the port? In other words, is it beneficial for the port and the city?* If we consider such developments an urban regeneration, four main elements should be addressed to change: economic, physical, social and environmental.

2.4 Emergence of Port-Regions as Growth Poles

The era of an increasing spatial requirements of technological advancement calls for port expansion and in some cases port relocation in post-industrial port-cities throughout the world (Hoyle and Pinder 1981). In the literature on regional planning and regional development studies, the 'growth pole' theory is often applied to new port infrastructures, in order to analyse and plan the developments occurred along the corridors, connecting the new and/or expanding port related facilities to the city centre (Dawson 1996; Hoyle 1996; Richardson 1978).

Recalling Bird's (1963) Anyport model in conceptualizing port development, three major steps were identified: setting, expansion and specialization. However, as argued by Notteboom and Rodrigue (2005), the model is weak in describing the contemporary port development since the model does not have the capacity to explain the emerging seaport terminals, such as transhipment hubs with limited or no local hinterland. Moreover, the traditional spatial model does not consider the inland freight distribution centres[4] and terminals as driving forces in port development dynamics. The scholars therefore improved the *Anyport model*: in addition to the existing three stages, a fourth phase is added: a displacement of port function into the region, called *port regionalization*. This concept introduces a broader geographical scale in port development perspective, enlarging the port hinterland through market strategies and policies, while providing closer connection to inland freight distribution cores. This stage of port development is where "*inland distribution becomes of foremost importance in port competition, favoring the emergence of transport corridors and logistics poles. The port itself was not the chief motivator for and instigator of regionalization. Regionalization resulted from logistics decisions and subsequent actions of shippers and third-party logistics providers*" (Notteboom and Rodrigue 2005: 311).

The port regionalisation concept is therefore characterized by deep functional interdependency of a load centre with a multimodal logistics base, which creates a

---

[4]'Inland port is a rail or a barge terminal that is linked to a maritime terminal with regular inland transport services. An inland port has a level of integration with the maritime terminal and supports a more efficient access to the inland market both for inbound and outbound traffic. This implies an array of related logistical activities linked with the terminal, such as distribution centers, depots for containers and chassis, warehouses and logistical service providers' (Rodrigue et al. 2013: 153).

'regional load network'. The emergence of this concept can be explained through two main factors:

- *Local issues*: Port developments are facing some local obstacles, such as the lack of land for their infrastructural expansion and the need for deep-water to handle larger ships. Port specialization, which increases the level of port traffic while creating diseconomies and environmental issues, mitigates such local constraints by externalizing them.
- *Global restructuring*: Globalization trends have substantially reformed the distribution system with the creation of regional production centres and large consumption markets. Thus, such a complex global supply chain cannot be handled by single localities. For instance, port regionalization gives rise to the globally integrated free trade zone (see Chap. 4) close to many load centres.

Port regionalization is also supported by the growth pole theory—the conventional location theory (Parr 1973; Perroux 1950) and the creation of logistics poles. The main idea behind the growth pole theory refers to the economic development that is not uniform and occurs around a specific pole, which is often composed of core (key) industries (such as automotive, aeronautical, agribusiness, electronics, steel, petrochemical, etc.) linked to secondary industries and contributes to economic development through twofold effects (Rodrigue et al. 2013):

- *Direct effects*: the core industry providing goods and services for its linked industries.
- *Indirect effects*: refers to the goods and services demanded by those employed in the core and linked industries.

Overall, the expansion in core industries may attract investments (including foreign direct investment—FDI), give path to new technologies and other industries to the hinterland, and therefore generate new jobs and employment. The scholars who consider ports as economic engines, define such infrastructures within the theory of growth pole, through economies of scale for production and trade; hence, benefiting the hosting regions and cities (Clark et al. 2004; Fujita et al. 1999). It is evident that seaports, serving as terminals, provide the site for transhipment to transfer commodities from one mode of transport to another, linking the maritime and inland transportation. Besides, the transferring activities may also necessitate a certain type of manufacturing to locate in close distance to the port. For example, the raw materials imported by ports are an opportunity for developing some processing industries. With globalization, the link between ports and transferring activities gave rise to about the creation of *logistics zones*. Therefore, within the concept of port regionalization, maritime terminals are considered as clusters and growth poles. The seaport clusters with concentrated interdependent organization, as defined by geographers and economists, are crucial in generating competition between regions and industries (Rodrigue et al. 2013).

In spatial terms, port regionalization moves the traditional attention on the role of ports at the local level to the complex reality at a broader and regional scale. From a political point of view, once again there is a shift of institutional power to the regional level; considering also the involvement of many actors managed by various entities. For geographer, port region is a broad term, yet not fully defined, which is a multifaceted concept containing the economic zone surrounding the port, the port hinterland (the logistics and free trade zone) and the broader port-system (see Chap. 3) (Ducruet 2009). Some scholars have made attempts in categorizing global port-region clusters, considering traffic volumes, types, and local economic structures across 1500 ports within 300 regions and 40 counties (Ducruet et al. 2012); an eight-typology classification is provided as follows:

1. *Deprived port regions*: characterised by high unemployment rate; specialized in handling liquid bulk; handling a small share of total world container traffic; insufficient industrial infrastructures and activities; limited economic growth that is mainly specialized in primary sectors.
2. *Peripheral port regions*: characterised by high unemployment rate; specialization in handling liquid bulk; handling a small of total world container traffic; insufficient industrial infrastructures and activities; high level of importing goods and service for local use.
3. *Metropolitan port regions*: own a high share of the national population, economic performance and port activities; high share in total international traffic (container and liquid bulk handling); service-based economy with limited productive activities.
4. *Industrial port regions*: own a high share of the national population, economic performance and port activities; high share in total international traffic (container and liquid bulk handling); secondary-based economy (production) prevails.
5. *Productive port regions*: characterised by high level of GDP; lower share of traffic than the world average; rich industrialized region (higher than average).
6. *Bulky port regions*: characterised by high level of GDP; lower share of traffic than the world average; large share of world container traffic (mainly solid bulk).
7. *Transit port regions*: lower level of liquid bulk traffic; higher valued goods and industrial activities dominates.
8. *Traditional port regions*: low population density; lower level of liquid bulk traffic; specialised in the primary sector (agricultural and mining).

Traffic flows in varied regions is highly correlated with the local economies (advanced, advancing, underdeveloped, etc.) and the socio-economic context of the port-region. Therefore, port development, port specialization and increase in traffic flows can only become a driver for economic growth considering certain local conditions.

2.5 Conclusion

Central to the topic of the book, this chapter has preliminarily focused on the changing relationship between the port and the city, while reviewing some general models describing the evolution of port-city interface in the European and Asian port-cities.

The port-city development worldwide is influenced by many landward and seaward locational factors. The original port has often determined the general layout of the city and hence an expansion of the port affects the spatial urban pattern. An essential common factor among port-cities worldwide, which determines the origins and settlement of spatial and socio-economic patterns, is the port function—transfer of goods—at the port-city interface. The general contemporary trend worldwide has been the relocation of port activates which has restructured the relation between the port and the city, both spatially and functionally.

Most port-cities today are or will eventually reach a point of becoming land constrained as the city continues to grow around it. Advancements in shipping and logistics technology and the eternal quest for economies of scale, has also led to a common demand element: the land. While technology can overcome the growing demand by processing more tonnage and containers from the same footprint, at some point additional land banks are required. Additionally, containerisation and the increasing automation of terminal operations have resulted in a decline in the overall complement of port workers. This has significantly changed most ports from being centres of major direct employment to centres of high technology logistics distribution, where the direct employment numbers, no longer have the same community impact as they once did. As discussed in this chapter, a broad range of disciplines has studied the port-city development issue from varied perspectives. Although, historically, ports and cities were strongly inter-linked, it seems that the current trend in major port-cities (especially in the Western world) is the weakening connection between the two elements. This is also due to the growing globalization and regional spill-over of the economic benefits, while negative impacts are localized (Merk 2010).

References

Akhavan M (2017) Development dynamics of port-cities interface in the Arab Middle Eastern world—the case of Dubai global hub port-city. Cities 60(part A):343–352. https://doi.org/10.1016/j.cities.2016.10.009

Akhavan M (2019) Contemporary European port-cities as laboratories. Territorio 88:99–104. https://doi.org/10.3280/TR2019-088015

Bird J (1963) The major seaports of the United Kingdom. Hutchinson, London

Bird J (1968) Seaport gateways of Australia. Oxford University Press, London

Bird JH (1971) Seaports and seaport terminals. Hutchinson, London

Bonfantini GB (2015) Historic urbanscapes for tomorrow, two Italian cases: Genoa and Bologna. Eur Spat Res Policy 22(2):57–71

Breen A, Rigby D (1996) The new waterfront: a worldwide urban success story. McGraw-Hill

Bruttomesso R (1993) Waterfronts: a new frontier for cities on water. International Centre Cities on Water, Venice

Bryan J, Munday M, Pickernell D, Roberts A (2006) Assessing the economic significance of port activity: evidence from ABP operations in industrial South Wales. Marit Policy Manag 33(4):371–386

Castro JV, Millán PC (1998) Port economic impact: methodologies and application to the port of Santander. Int J Transp Econ/Riv Int Econ Trasp 25(2):159–179. https://doi.org/10.2307/42747269

Clark X, Dollar D, Micco A (2004) Port efficiency, maritime transport costs, and bilateral trade. J Dev Econ 75(2):417–450. https://doi.org/10.1016/j.jdeveco.2004.06.005

Daamen TA, Louw E (2016) The challenge of the Dutch port-city interface. Tijdsch Econ Soc Geogr 107(5):642–651. https://doi.org/10.1111/tesg.12219

Danielis R, Gregori T (2013) An input-output-based methodology to estimate the economic role of a port: the case of the port system of the Friuli Venezia Giulia Region, Italy. Marit Econ Logist 15(2):222–255

Dawson AH (1996) Cityport development and regional change: lessons from the clyde. In: Hoyle B (ed) Cityports, coastal zones, and regional change: international perspectives on planning and management. Wiley, pp 49–57

Ducruet C (2006) Port-city relationships in Europe and Asia. J Int Logist Trade 4(2):13–36

Ducruet C (2009) Port regions and globalization. In: Notteboom TE, Ducruet C, de Langen P (eds) Ports in proximity: competition and coordination among adjacent seaports. Ashgate, Aldershot, pp 41–53

Ducruet C, Jeong O (2005) European port-city interface and its Asian application. Research report 17. Korea Research Institute for Human, Anyang

Ducruet C, Itoh H, Joly O (2012) Port-region linkages in a global perspective. In: MoLos conference modeling logistics systems, pp 1–25

Fujita M, Mori T (1996) The role of ports in the making of major cities: self-agglomeration and hub-effect. J Dev Econ 49:93–120

Fujita M, Krugman PR, Venables A (1999) The spatial economy: cities, regions, and international trade. MIT Press, Cambridge

Gleave MB (1997) Port activities and the spatial structure of cities: the case of Freetown, Sierra Leone. J Transp Geogr 5(4):257–275. https://doi.org/10.1016/S0966-6923(97)00022-7

Gripaios P, Gripaios R (1995) The impact of a port on its local economy: the case of Plymouth. Marit Policy Manag 22(1):13–23. https://doi.org/10.1080/03088839500000029

Haezendonck E, Coeck C, Verbeke A (2000) The competitive position of seaports: introduction of the value added concept. IJME 2(2):107–118

Hall P, Jacobs W (2012) Why are maritime ports (still) urban, and why should policy-makers care? Marit Policy Manag 39(2):189–206

Hayut Y (1981) Containerization and the load center concept. Econ Geogr 57(2):160–176

Hayuth Y (2007) Globalisation and the port-urban interface: conflicts and opportunities. In: Wang JJ, Olivier D, Notteboom T, Slack B (eds) Ports, cities, and global supply chains. Ashgate, Aldershot, pp 141–156

Hein C (2011) Port cities: dynamic landscapes and global networks. Routledge, London

Hesse M (2018) Approaching the relational nature of the port-city interface in Europe: Ties and tensions between seaports and the urban. Tijdschrift Voor Economische En Sociale Geografie 109(2):210–223. https://doi.org/10.1111/tesg.12282

Hesse M, Rodrigue J-P (2004) The transport geography of logistics and freight distribution. J Transp Geogr 12(3):171–184. https://doi.org/10.1016/j.jtrangeo.2003.12.004

Hilling D (1977) The evolution of a port system—the case of Ghana. Geography 62(2):97–105

Hoyle BS (1968) East African seaports: an application of the concept of "anyport." Trans Inst Br Geogr 44:163–183

Hoyle BS (1983) Seaports and development: the experience of Kenya and Tanzania. Gordon and Breach, New York and London

Hoyle BS (1988) Development dynamics at the port-city interface. In: Hoyle BS, Pinder D, Husain S (eds) Revitalising the waterfront: international dimensions of dockland redevelopment. Belhaven Press, Great Britain, pp 3–19

Hoyle BS (1989) The port—city interface: trends, problems and examples. Geoforum 20(4):429–435. https://doi.org/10.1016/0016-7185(89)90026-2

Hoyle BS (1993) Some Canadian dimensions of waterfront redevelopment. In: Bruttomesso R (ed) Waterfronts—a new frontier for cities on water. International Center Cities on Water, pp 333–338

Hoyle BS (1996) (ed) Cityports, coastal zones, and regional change: international perspectives on planning and management. Wiley, Chichester

Hoyle BS (2000) Global and local change on the port-city waterfront. Geogr Rev 90(3):395–417. https://doi.org/10.1111/j.1931-0846.2000.tb00344.x

Hoyle BS, Hilling D (1970) Seaports and development in tropical Africa. Macmillan, London

Hoyle BS, Pinder D (1981) Cityport industrialization and regional development: spatial analysis and planning strategies. Pergamon Press, Oxford

Krugman PR (1997) Development, geography, and economic theory. MIT Press, London

Lee S, Ducruet C (2009) Spatial glocalization in Asian hub port cities: a comparison of Hong Kong and Singapore. Urban Geogr 30(2):162–184

Lee S-W, Song D-W, Ducruet C (2008) A tale of Asia's world ports: the spatial evolution in global hub port cities. Geoforum 39(1):372–385. https://doi.org/10.1016/j.geoforum.2007.07.010

Marshall R (2004) Waterfronts in post-industrial cities. Taylor & Francis, New York

Merk O (2010) The competitiveness of global port-cities: synthesis report, p 184

Musso E, Benacchio M, Ferrari C (2000) Ports and employment in port cities. IJME 2(4):283–311

Notteboom T, Rodrigue J (2005) Port regionalization: towards a new phase in port development. Marit Policy Manag 32(3):297–313

Notteboom TE, Winkelmans W (2001) Structural changes in logistics: how will port authorities face the challenge? Marit Policy Manag 28(1):71–89. https://doi.org/10.1080/03088830119197

O'Connor K (2010) Global city regions and the location of logistics activity. J Transp Geogr 18(3):354–362. https://doi.org/10.1016/j.jtrangeo.2009.06.015

Omiunu FGI (1989) The port factor in the growth and decline of Warri and Sapele townships in the western Niger Delta region of Nigeria. Appl Geogr 9(1):57–69. https://doi.org/10.1016/0143-6228(89)90005-2

Parr JB (1973) Growth poles, regional development, and central place theory. Pap Reg Sci Assoc 31(1):173–212. https://doi.org/10.1007/BF01943249

Perroux F (1950) Economic space: theory and applications. Q J Econ 64:89–104

Richardson HW (1978) Regional and urban economics. Penguin, Harmondsworth

Rodrigue J-P, Comtois C, Slack B (2013) The geography of transport systems, 3rd edn. Routledge, Abingdon

Slack B, Frémont A (2005) Transformation of port terminal operations: from the local to the global. Transp Rev 25(1):117–130. https://doi.org/10.1080/0144164042000206051

Storper M (1997) The regional world: territorial development in a global economy. Guilford Press, New York

Van den Berghe K, Jacobs W, Boelens L (2018) The relational geometry of the port-city interface: case studies of Amsterdam, the Netherlands, and Ghent, Belgium. J Transp Geogr 70:55–63. https://doi.org/10.1016/j.jtrangeo.2018.05.013

Wang JJ (2014) Port-city interplays in China. Ashgate, Farnham

Chapter 3
Revisiting Port-Cities in the Global Context

3.1 The Notion of Globalization and World/Global City Theory

Back in the mid-1960s, Sir Peter Hall claimed that: 'There are certain great cities, in which a quite disproportionate part of the world's most important business is conducted' (Hall 1966: 7). Twenty years later and even today, globalization trends and the world/global city theory has been the subject matter of many studies and discussions (Castells 1989, 1996; Friedmann 1986; Knox and Taylor 1995; Sassen 1991, 1994). Although globalization of the world-system, through trade and ports, goes back more than a century, the shift from an international to a global economy has started during the 1970s and 80s (Knox and Taylor 1995). Considering the close link of urbanization and the global economic forces, some scholars argue that a city's position within a network—of people, goods, capital and information—affects its economic performance and competitiveness at the regional and global scale (Amin and Thrift 2002; Marcuse and Van Kempen 2000; Smith 2001).

In his world city hypothesis, Friedmann (1986) studies the relation between urban networks and world economies. He argues that within the world city system, key cities are emerging as 'basing points' for global capital, which may result in a complex spatial hierarchy. While it is discussed that the command and control functions of the global economy are concentrated in only few cities at the top of the urban hierarchy, major world/global cities are due to acquire new roles and functions, emerging from spatial dispersion and global integration (Sassen 1991, 1994).

Cities respond differently to globalization; hence, the level of global competitiveness and connectivity also differs in urban centres. To understand the nature of this global urban hierarchy, a number of scholars have made attempts to measure the level of 'world/global city-ness', through using a set of criteria to rank the cities. The pioneering works by Hall (1966), Hymer (1972) and Heenan (1977) analysed the role of multinational cooperation (hereinafter MNC) headquarters with a focus on the Western Developed world. Hall (1966) described the cities on the top urban hierarchy to have global functional capacities, with power in finance, trade, communication, education and technology. While according to Hymer (1972), most MNC headquarters are concentrated in major cities, as they seek to locate in close proximity to the capital market, the media and the government.

An important contribution into the global urban hierarchy literature is the studies carried out by Friedmann (1986), which described major cities as control centres of capital within the new international division of labour. He analysed his world hierarchy based on seven key factors: (i) major financial centres; (ii) MNC headquarters; (iii) international institutions; (iv) rapid growth in business services sectors; (v) important manufacturing centres; (vi) significant transportation nodes; and (vii) population size. Another approach by Reed (1981) suggested the analysis of international financial centres as the basis to characterize and position the major cities. And later on, during the 90s and onward, one of the most influential scholars in the world/global city studies is definitely Saskia Sassen, which discusses an urban hierarchy with relation to internationalization, agglomeration and concentration of producer services—namely finance, insurance, accountancy, advertising, banking, etc.—in the new urban economy (Sassen 1991, 1994). As key sites for specialized service firms, production and innovation, she argues that the major global cities on the top list—New York, London and Tokyo—function as the key command nodes within the system of global economy.

Taking advantage of a large quantity of comparative data, Beaverstock et al. 1999 introduced the Globalization and World Cities—GaWC[1] inventory of world cities based on their provision of global corporate service firms in all advanced producer service (APS) sectors. The authors have identified and classified world cities as global service centres based on the level of accountancy, banking/finance and law. Although their research is also based on APS, in which the city is considered as a global financial/service centre, however compared to the previous studies, here a wider classification is introduced and therefore a relatively large number of cities are considered world cities at different levels. In fact, since 1998, the GaWC Research Network have extensively studied the impact of world cities in globalization trends, while elaborating a geographic and economic-based overview of the world's evolving configuration. Based on their international connectedness, the cities are classified into four main groups: Alpha (α), Beta (β), Gamma (γ) and Sufficient[2] (Taylor 2004).

Figure 3.1 shows the map of the world city connectivity according to GaWC (2010). The map clearly illustrates some parts of the world being rather dense in contrast to the less connected areas, in terms of world city connectivity: Western World is far more connected, which also confirms the stark distinction described by the North-South divide. Although the Global south demonstrates a relatively poor connection within the whole picture, some red and green circles (Alpha and Beta

---

[1] GaWC stands for the Globalization and World Cities Research Group, which is a research network based in the Geography Department at Loughborough University (UK), founded by Taylor in 1998, where they study the relationships between world cities in the context of globalization. https://www.lboro.ac.uk.

[2] The classification is based upon the office networks of 100 advanced producer service firms in 315 cities. Cf. GaWC Research Bulletin 43. https://www.lboro.ac.uk/gawc/world2000.html.

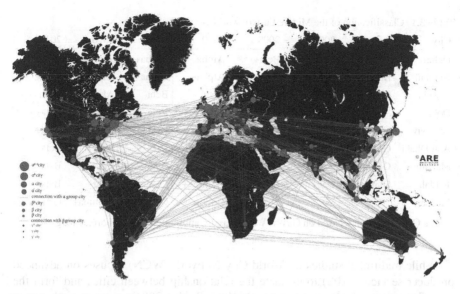

Fig. 3.1 Map of global cities, according to GaWC in 2010. *Source* Borrowed from the GaWC https://www.lboro.ac.uk/gawc/

cities) are also emerging. Relevant to core case study of this book, which is the case of Dubai (see Chaps. 4 and 5), it is worth mentioning that according to the latest study by GaWC (year 2018), Dubai is ranked as an Alpha+ city, together with sever other world cities: Hong King, Beijing, Singapore, Shanghai, Sydney, Paris and Tokyo. In fact, since 1991, Dubai has been the first Arab Middle Eastern city to feature a world city connectivity (Shin and Timberlake 2000); and since 2010 it appears as top important cities within the global network. Although Dubai is well connected, mainly thanks to the office networks of advanced producer services, yet according to the study by Bassens (2013), it is not strategically important; he simply refers to Dubai as a 'gateway global city' for its specific region. Table 3.1 shows the emerging world cities of classification in the Middle East—since 2010 till 2018; all cities in the list host a port: thanks to the growing importance of the Arab States port-cities, this part of the port is becoming more and more connected to the world network (see Chap. 4).

Verhetsel and Sel (2009) have applied the GaWC methodology to explore the world maritime city network based on the presence of container shipping companies and terminal operators, and their degree of connection between these sectors and the world's maritime market. This study confirms that many world cities are also port-cities—i.e. have maritime base—as historically important nodes of transportation within the world maritime city network: Hong Kong, Hamburg and New York are the main nodes; although London has no longer an important port, it is ranked as Alpha World Maritime City. Relevant to the focus of this book on the port-cities of the Eastern World: Singapore and Dubai are also classified as leading centres, respectively an Alpha and Beta World Maritime City.

Table 3.1 Classification of the Middle Eastern world cities

City	2000	2004	2008	2010	2012	2016	2018
Dubai	Beta	Beta	Beta+	Alpha	Alpha+	Alpha+	Alpha+
Istanbul	Beta+	Beta	Alpha−	Alpha−	Alpha−	Alpha	Alpha
Riyadh	Gamma	Gamma−	Beta	Beta	Beta	Alpha−	Alpha−
Doha			Gamma−	Gamma+	Beta	Beta	Beta+
Tel Aviv	Gamma	Gamma+	Beta+	Beta+	Beta+	Alpha−	Beta+
Abu Dhabi	Gamma−	Gamma		Beta−	Beta−	Beta	Beta
Manama	Gamma		Gamma+	Beta−	Beta−	Beta	Beta
Jeddah	Gamma	Gamma−	Beta−	Gamma+	Gamma+	Beta−	Beta−
Muscat				Gamma−	Gamma	Gamma+	Gamma+

Source Own elaboration based on GaWC World classification https://www.lboro.ac.uk/gawc/

While traditional studies on World City Network (WCN) focuses on advanced producer services (APS) to evaluate the relationship between cities and form the intra-firm networks (Beaverstock et al. 1999; Taylor 2004), more recently, other scholars have expanded the research to include other services and industries such as the maritime and port-related APS: Jacobs et al. (2010) have used the maritime APS firms (such as maritime law, P&I clubs, insurance brokers, classification societies, consultancy, surveyors and maritime organisations) to investigate the role of port-cities in global commodity systems and the world city network. Their findings highlight that, in general, the maritime APS tend to follow the world city hierarchy, with some port-cities acting as main nodes in global commodity chains. Although the urban economic impact of being a port-related APS centre was not left unanswered, it is stated that the relatively strong position of Dubai is associated with its rapid urban economic development in a short time span of two decades.

The studies that have focused on understanding the interaction between globalization and cities, highlight the challenge in conceptualizing a city's level of integration when positioning cities within the world network. Despite this acknowledgeable effort, there is still a gap in the literature in providing a comprehensive understanding of how each city functions within the global economy (Parnell and Robinson 2012). The focus of the reviewed literature has, therefore, been mainly on defining the cities that are affected by globalization and measuring their level of *global-ness*. There is the need to study not only how cities are affected by the globalization but also to understand how they respond to the forces.

The globalization and increasing level of trade, merchandise and goods has led to the re-territorialization of socio-economic and spatial spaces that are at the conjunction point of the overlapping scales. In this study, the case of port-cities is foreseen as laboratories for a productive approach in transnational new urbanism. The notion of transnational urbanism acts at the multiple yet overlapped spatial scales thus can be defined as mutual interaction of flows at local, regional, national and transnational level. In this case the local is redefined, as an outcome of multiple interaction of overlapping scales (Smith 2001).

3.2 Ports as Integrated Elements in Logistics and Freight Distribution

The *early modern globalisation*—between the years 1600 and 1800—is characterised by increasing trade links and cultural exchanges that was immediately followed by the so-called 'modern globalization' of the late nineteenth century (Hopkins 2003). The increasing connection between the East and West in this era, also known as *proto-globalization*, was marked by the rising maritime European Empires of the fifteenth century—mainly the Portuguese, the Spanish, the Dutch and the British Empires. In such expansion of trade routes and transportation, maritime ports played a key role; and this is why in the phase of early modern globalisation many port-cities were developed as important urban centres that continue to thrive even today. The shift to modern globalisation (the one described in the previous section)—as a complex process of worldwide integration, interaction and flows of people, goods, capital and power—has led to a substantial increase in the volume of cargo, which necessitates a more sophisticated trading system; many ports were therefore forced to adapt themselves to the changes in order to be inserted into the wider network of logistics and the supply chains.

Logistics is defined as the process of planning, implementing and controlling the flows and/or stored goods, services and information from the point of origin to the final destination (consumption) (Coyle et al. 1999). The supply chain is a wider term, and according to the Council of Supply Chain Management Professionals (2007), the supply chain management encompasses all logistics activities with an emphasis on the co-operative and coordinative role of different actors—suppliers and end users— to facilitate the most efficient flows of goods and information. Accordingly, ports are not only based at the point to facilitate the receipt and dispatch flows of commodities, information, and related processes; but also within a supply chain complex ports are involved more actively on how activities and processes beyond its borders may have impacts on the value added and supply chain management (Panayides and Song 2008).

As a central and fundamental feature of economic activity, logistics process throughout the process of production, distribution and consumption has been the subject matter of many studies (see Mariotti 2015). For Hall and Hesse (2013), logistics is fundamental to the relationship between spaces and flows. Others have discussed the complex interdependent commodity distribution system and coordination of flows, while underlining that since the revolution in shipping and transportation industry massive container traffic are handled by major terminals of infrastructural nodes, such as seaports and airports (Hesse and Rodrigue 2004). Logistics related activities are therefore to locate in close proximity to major gateway and hubs where they have access to a relevant market area; also to agglomerate in hinterland corridors, even beyond city boarders (Hesse 2004; O'Connor 2010; Holl and Mariotti 2017).

Here it is important to define which logistics functions are truly port related. In fact, all activities related to the berthing of vessels, and its loading and unloading, also activities at the container terminal (cargo receiving, handling and delivering) which necessitates the proximity to the port terminal, and are performed by various actors of port logistics, such as container terminal and port authorities, exporters, importers, shipping industries, freight forwarders, custom brokers, etc. In other words, in new logistics environment, logistics activities can be categorized as:

- Activities that lead to relevant decrease in transported volume;
- Activities related to big volumes of bulk cargoes, relevant to inland transport and rail;
- Activities that are directly linked to industries and companies owning a site in the port zone;
- Activities related to cargo handling that require specific storage buffer;
- Activities related to high demands for short-sea shipping (Notteboom and Rodrigue 2005: 18).

Ports integrating as important elements within the global supply chain can be traced back to the advent of containerization, which led to the development of modern port infrastructures. As discussed beforehand the two models of Bird (1963) and Hoyle (1989) have considered the phase in which the port faces a major geographical and technical transformation to host large container ships. Nevertheless, traditional research on ports were focused on more physical and functional aspects of the port at the city level, but also studies on the port as a node within the transport regional and global network (Olivier and Slack 2006). Recent studies, however, are showing a growing interest on the role of ports within the advanced system of transportation, logistics system and the global supply chain, from which emerged the concept of port-centric logistics and port regionalization (Bichou and Gray 2004; Carbone and Martino 2003; Heaver 2002; Mangan et al. 2008; Notteboom and Rodrigue 2005; Panayides and Song 2008; Robinson 2002).

Globalization of economic, market, production and distribution have led to an evolution in trade and activities of logistics operators. Starting from the early twenty-first century, scholars have highlighted the uncertainty and challenges of the ports' management and port authorities in the modern logistics system (Notteboom and Winkelmans 2001) and the shortage of the existing paradigms to explain the functions of the port within the global supply chain. In this regard, Robinson (2002) introduced a new paradigm by defining the role of port through its function as an element within the new logistics system and value-added supply chains. Although ports are inserted within a newly competitive globalized, corporatized and privatized environment, they are also part of a logistics-restructured base. This advantages major players, such as shipping lines, port-authorities and logistics service providers to have an important role in controlling the logistics chain. Therefore, the port becomes (i) a third-party service provider as part of individual firm's supply chain; (ii) an element (or firm) along with other firms associated to the import and export supply chain; (iii) a provider of superior value delivery to shippers; (iv) a rival to other ports (as a firm) within a port-related supply chain.

Furthermore, ports are increasingly becoming important elements for maintaining the regional and global competitiveness. In defining port characteristic, the port efficiency is considered among the most important factors, including port infrastructure, private sector participation and inter-port connectivity. Some scholars have studied the relation between port efficiency, transportation costs and level of trade, affecting the regional and national competitiveness. The study by Cullinane and Song (2002) shows the port's critical role and the level of efficiency and performance in the supply chain being closely linked to the national competitiveness. Moreover, in the context of Latin American countries, Sanchez et al. (2003) underlined that while port efficiency is an important factor in reducing the cost of trade, it is also a relevant determinant of country's level of competitiveness.

For Jacobs and Hall (2007) ports are territorially embedded in a historically path dependent institutional system. The authors use the concept of insertion, integration and dominance to argue that a port's level of *embedded-ness* in the global supply chain and the strategies taken by major port actors are conditioned by the nature of the port territorial embedded-ness. The concept is further expanded through studying the case of Dubai as an example of a local port authority transformed successfully into a global terminal operator (for more detailed reading on Dubai-based global operator DP World see Chap. 5). In this specific case, the port authority has a full control over the port hinterland the infrastructures; nonetheless, the State as a multinational corporation is an important player and integrates its horizontal and vertical organizational functions to develop economic and spatial synergy.

Some studies have expanded the World City Network (WCN) by GaWC (discussed in the previous section) to include the advanced logistics services (ALS)—such as third-party logistics—3PL and fourth-party logistics 4PL provider (Antoine et al. 2017; O'Connor 2010; O'Connor et al. 2016; Wang and Cheng 2010); it is therefore discussed that ALS are crucial in the control of spatial and information flows in the global economy (Hesse and Rodrigue 2006; Jacobs et al. 2010). Recently, some scholars have developed an interlocking network of ALS in Europe and shows that 3PL management functions (that generate logistics connectivity) are attracted by: (a) World Cities as knowledge-rich environments and connected places, and characterized by urbanization economies, and (b) port and airport infrastructures as clusters of specialized knowledge (transport and logistics industry), where localization economies can be exploited' (Akhavan et al. 2020b). Another similar study has applied the same method to examine the attractiveness of European maritime port-cities to the largest global 3PL providers (Akhavan et al. 2020a): due to the effect of specialized transport centres, even port-cities that are geographically distanced from world cities can attract ALS activities; on the other hand, some world cities that are in close proximity to main ports may become inland ports (such as the case of Milan with respect to the maritime ports of the Liguria (Genoa, La Spezia and Savona), and therefore sucks up the ALS functions. This effect causes the port-city to lose its value-added sectors and employment, which may limit the local economic growth.

3.3   The Significance of Free Trade Zones Within the Port Hinterland

According to United Nation ESCAP, port-hinterland (or just hinterland) is: *"the land area located in the vicinity of the port such as immediately nearby or within the port boundary and functioning interactively and closely with a port by providing various business activities, whether or not the hinterland is within the administrative jurisdiction of the port authority"* (United Nations 2005: 16). Other scholars define port-hinterland as the inner region provided by a port, where the port has a monopolistic role (Fageda 2005). Accordingly, two main features can be highlighted: (i) the proximity of the area to port terminal operations, and (ii) the accessibility of the hinterland to port's clients.

Tan (2007) categorizes the typologies of such areas as the immediate hinterlands (ports area), primary hinterlands (where port and city have a controlling position), commodity hinterland (relate to the type of commodities), and the inferred hinterland (port's domination over a particular area in order to meet the demands for imports). Given that the transport and transhipment activities need certain types of manufacturing operations to locate close to the port, it may generate new employment, economic investments and activities, as well as the designation of special tax and economic zones, such as Free Trade Zones (hereinafter FTZ). Port's performance and competitiveness is associated to its relationship with the hinterland (Danielis and Gregori 2013). In order to respond to global changes, as well as to enter the global supply chain, many maritime ports have developed logistics clusters in their hinterland. The agglomeration of logistics firms in port area provides collective advantages that may not be easy to access in separate locations. The port-centric logistics zones in the port-hinterland can, therefore, be classified as follows (Chen and Notteboom 2012):

- *Traditional port-based logistics zone*, which is related to the pre-container port area.
- *Container based logistics zone*, which is the most dominant kind with numerous warehouses in close vicinity to the container port and intermodal terminal facilities.
- *Specialized port-based logistic zone*, which can be specialized according to different functions related to the characteristic of the port.
- *Logistics-oriented Free Trade Zone* (hereinafter FTZ).

The phenomena of FTZ are emerging clusters of multi-functional firms inside the port area or in close proximity to the maritime port and/or airport, *'where goods can be imported, stored or processed and re-exported, free of all Custom Duties. It is a free area which normally falls under the authority of the port or airport management'* (UNCTAD 1996: 3). Here it is worth underlining that the history of FTZ is tied to the advent of free ports (or freeports), which have existed for centuries. It basically covers a much larger area (in some cases it includes the entire port-city) and can host a wide range of activities, incentives and benefits to boost economic development and trade (World Bank 2008). As key elements of the trade system, contemporary free ports have been subject to change in terms of their role and function to adapt to the political, economic and technological development. Lavissière and Rodrigue (2017) discuss three main factors that may explain the evolution of free ports: the regulatory context, the function and trade orientation. FTZs are only one kind of the broader term Special Economic Zones (SEZ), with specific rules governing local economic activities, which are different to the rest of the country; the aim is to attract capital, investment, production and companies. Apart from FTZ, the Export Processing Zones (EPZ), the High-Tech Development Zones, and the Science Technology Parks, are other typologies of SEZ (Barbieri and Pollio 2015; Easterling 2014).

Since the 1970s, FTZs are developed as a rather successful policy instrument to promote export-oriented foreign direct investment (hereinafter FDI), which have been widely applied in both developed and developing countries (United Nations 2005). In general, trade and FDI flows are popular among academics and practitioners as important factors in encouraging economic growth in host counties (Castellani and Zanfei 2006; Mariotti and Piscitello 2006). Inward FDI can expand the host country's trade capacity, developing technology, creating new jobs and boosting an overall economic growth in host counties (Balasubramanyam et al. 1996; Lipsey 2000). The literature on FDIs indicates that Multinational Enterprises (MNEs) allocate their investments among countries to maximize their risk-adjusted profits (Caves 1996). Regarding the "Location" group of factors, infrastructures play a key role in attracting FDI, being them either transport infrastructures, such as ports, or their hinterland development, namely FTZ (Akhavan and Mariotti 2015). In fact, the literature analysing the determinants of inward FDI towards the hinterland in regional ports has underlined the following location factors: government supporting policy, FTZ, industry cluster and ICT infrastructure (Cho and Ha 2009). Generally speaking, the role of infrastructure development as an engine in attracting FDI is widely studied and accepted by scholars. Nevertheless, compared to the other modes of transport infrastructures, the studies on the determinants of inward FDI flows towards maritime ports and its hinterland are less rich. Apart from a limited investigation, through case-study approach in developing countries, (Belloumi 2014; Cho and Ha 2009), there is a lack of comprehensive study on FDI attraction in cities with major container ports.

Given the economic basis of the FTZ, developing countries can take advantage of this concept towards economic diversification and entering the world economy, while deploying a neoliberal reform through weakening state's power over the territory and policy making (Papadopoulos 1987). In urban studies, the debate on FTZs are often centred on the strong global pressures toward liberalization of local markets, and how the immobility of the state's territorial control leads to a recurring cycle of de- and re-territorialisation (Keshavarzian 2010). Furthermore, in the same filed, some scholars refer to these zones as extrastatecraft—a portmanteau meaning both outside of and in addition to statecraft—which goes beyond its traditional function as a remote locale for warehousing and transhipment, so becomes a 'city' (Easterling 2014).

3.4   The Concepts of Entrepôt, Gateways and Hub Port-Cities

Back in the early twentieth century, the term *entrepôt* was defined by Smith (1910) as a city that could successfully mediate the flows of traffic within a region. The industrial revolution of the nineteenth century in the more advanced capitalist countries was accompanied by the development of entrepôts, as resource processing, administrative and trading centres in Asia, Latin America and part of Africa, mainly under the power of colonial rulers (Scott and Storper 2014). The free ports of Singapore (Yeoh 2003), Hong Kong (Tsang 2007) and Dubai Creek (Akhavan 2017) can be referred to as important examples of the Eastern historical entrepôts. Since the colonial era, all the three examples are serving as international intermediaries, which distinguishes them from their regional counterparts.

Not to be confused with the French meaning of warehouse, an entrepôt is a city or a trading post, strategically situated at the conjunction point of major trading routes, where goods are imported, stored and/or re-exported (Muller 1976). Traditionally, the concept of an entrepôt has been mainly used to describe the role of cities within the mercantile world-system, generally for the case of historical colonial trading ports (Ken 1978). However, most recently Sigler (2013) has made an effort to conceptualize entrepôts (and gateway cities) as relational cities in the contemporary global production networks, which form an interface between global flows, networks and cities, in order to mediate the flows of regional re-exportation.

As a consequence of a rising global trade, advancement of the transport industry and restructuring global economy, and therefore the changing role of places the importance of world entrepôts were overshowed by the emergence of gateways and transportation hubs. In fact, from the mid-twentieth century, following the decolonization and emergence of the nation-state system, entrepôts studies became dominated by the emergence of the gateway cities. The concept of gateway cities has been well illustrated by Burghardt (1971), which defines it as an entrance into a larger hinterland. In contrast to the central place theory, gateway cities are created through trade and service areas, located at one end of a fan-shaped economic system instead of being constant central places. Short et al. (2000) define a gateway city as an economic transition zone, which is becoming as important as the nation-state; yet contributing to the rather limited literature on globalization, they propose this concept to be applied to all cities as transmission points.

Despite the importance of gateway cities for a region's prosperity, Hesse (2010) argues the competitive advantage of gateway cities that are challenged by issues such as adapting to the changing spatial conditions, related to the physical constraints, land use or access limitations, and also the relevant to locational dynamics. In the case of major port-cities with a gateway position they can benefit from the possibilities to attract value-added logistics and other activities that require ports proximity, in order to develop maritime clusters (Notteboom and Winkelmans 2001), which is defined as: 'the set of interdependent firms engaged in port related activities, located within the same port region and possibly with similar strategies leading to competitive advantage and characterised by a joint competitive position vis-à-vis the environment external to the cluster' (Haezendonck 2001).

Recalling one of the aims of this book, which is exploring the formation and significance of 'hub-port' cities, here specific attention is reserved to define hubs. In order to study the role of cities as 'hubs', one should consider them within a network of flows; due to the existence of varied type of networks, specialized hub-cities are emerging. On this matter, some scholars have used the concept of *'centrality'* along with other concepts to explain the attributes related to the hubs. Exploring the importance of locational features in terminals, Fleming and Hayuth (1994) have used the two concepts of *centrality* and *intermediacy* as two spatial qualities that determines the level of traffic in transportation hubs, and therefore defining places that are strategically located in the network. They argue that a port-city's level of centrality is substantially relevant to the size and function of that city; while intermediacy implies the spatial quality of the 'in-betweenness' of the specific context and transportation system and terminals.

To be more specific, from a spatial point of view, the concept of 'centrality' has been identified as city's geographical location in the middle (or centre) of a territory/region, which resembles the 'central place theory' (see Christaller and Baskin 1966). However, the notion of centrality for a transportation port situated in a network is related to a city's status at the centre of interactions and exchanges with other cities. Based on a detailed study by Freeman (1979), a city's centrality, as a node in a network can be defined in three distinct conceptions: (i) *degree centrality*, which focuses on each node's number of connections; (ii) *betweenness centrality*, which is equal to the number of shortest paths from all connections that passes through that node; in other words is about the significance of a node for moving from one part of a network to another (this is useful as an index to define the state of a point to control the communication); (iii) *closeness centrality*, which is concerned with either independence or efficiency, in which determines a node's accessibility to other nodes. Therefore, the degree centrality is a local measure; other two measures are relative to the rest of the network. Each of the three conceptualized node centrality implies a different perspective of network positioning that is useful in analysing the specific attributes of interactions.

In an attempt to examine centrality measures with respect to theoretical specifications of urban hierarchies, Irwin and Hughes (1992) have tested the applicability of *graph theory* to urban systems, associated with spatial economic theories (Fig. 3.2). Accordingly, they discuss the network position, in terms of centralized measures, is essential to the three fundamental theories of the (regional) economic developments:

1. *Economic base theory* (trade volume and centrality): based on this theory, the city's economic strength is highly dependent on its basic industries (vs. the non-basic[3]), which relies on the exporting activities. Cities with high level of exportation tend to interact with many other cities and hence become more dominant: such city is more likely to influence the trading market and the life of other

---

[3]In the economy, basic sectors are related to local businesses (firms) that are entirely dependent on external factors. Locally based resources such as mining and agriculture, manufacturing, and tourism

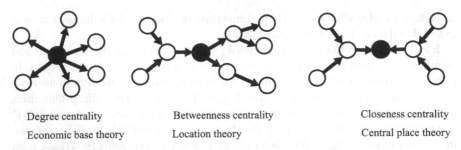

Degree centrality	Betweenness centrality	Closeness centrality
Economic base theory	Location theory	Central place theory

Fig. 3.2 The three measures of centrality versus the theories of regional economic development. *Source* Adapted from Irwin and Hughes (1992)

cities becomes more dependent on it (Alexander 1954; Andrews 1953). Hence, in this approach, centrality is parallel to the level of exportation (Clark 1945). In network positioning, the degree centrality is determined by direct activities; and then according to the economic base theory, the places with dominant direct connections become central places. This economic theory, hence, rests on the degree centrality, whereas the centrality is based on the volume of exportation.

2. *Location theory* (growth and development through mediating the exchange): this theory is preliminarily concerned with the optimal location of cities and place, balancing both the land and transportation cost (see von Thunen 1826). Based on this theory, cities located at certain points within the transportation systems (mainly at the middle of a supply chain), may have dominant economic potentials, serving as bridges, to mediate and coordinate the commodity flows among the places (Cooley 1894). In this case, the city becomes a mediator and facilitator of flows from one place to another, thus producing economic advantages for early and subsequent growth.[4] The centrality, here, is based on the number of flows passing through the city, as a connection point between cities, which are otherwise disconnected. Therefore, as the location theory stresses the city as mediator of flows, the betweenness centrality best fits its position within the network.

3. *Central place theory* (location and accessibility): this theory argues that certain centralized places (cities) which have maximum access to consumers are more likely to develop a diversified economy, through providing services and goods to the surrounding area (Christaller and Baskin 1966). In such network, the level of centrality is calculated through the number and volume of flows coming from other places (cities): main central places are attracting consumers from other places, whereas others draw fewer consumers from a limited number of places.

---

are mainly referred to as basic sectors as their existence depend largely upon non-local factors. Non-basic industries are, in contrast, defined as sectors that requires local business conditions, such as local services (shops, public schools, local government, etc.).

[4]Early stage of economic growth is derived from industries that facilitate transformation or distribution. Along with the city's growing central dominance, further development is occurred through rising imports and economic agglomeration (Irwin and Hughes 1992).

Since this theory is based on accessibility and relative centrality of other places, hence the closeness centrality is the suitable measure of network position. The conception of closeness centrality is relevant to the independence measures, which is parallel to the concept of central place theory. In other words, the most central places are independent from the other places, in terms of providing goods and services to the local market, as others are dependent to the major central place.

If cities are thought of as 'nodes', then the critical centres what mediate the flows in and out can be considered as 'hubs'; and based on this framework, new hub-cities are emerging. For port cities, 'hub' refers to ports with limited transhipment traffic (including gateways) or solely transhipment traffic, which is the case in both Singapore and Dubai. But the term 'transhipment hub port' is used to describe cases where transhipment traffic is more than half of the port's traffic (Nam and Song 2011).

The nature of hubs continues to change in a world of growing globalization and the intensification of activities that occur within port regions, which creates multilayer regional trading centres (Rimmer 1999). Wang and Cheng (2010) introduce a four-tier hierarchy for multi-player trading hubs in Asian mega-cities, like Singapore and Hong Kong. Their model defines a major global or regional trading hub based on four transport and logistics activity criteria: (1) maritime transport with a high level of shipment; (2) a well-linked hinterland supported by efficient land transport system; (3) an air transport hub or gateway; (4) a non-physical logistics platform to form a more efficient supply chain management system. The significance of each feature may change depending on the level of globalization, efficiency of global trade organizations, and the wealth of the hub city in comparison to the rest of the region.

According to Rimmer (1999), major hub port-cities are not only hubs based on the level of container turnover but have also been important centres for global trade since the 1980s. Considering the fact that 90% of world trade occurs via maritime transport (International Maritime Organization 2012), ports are therefore considered key aspects of the global value chain (Robinson 2002). Airports can also be considered as multi-scaled hubs in the global supply chain, as some studies have verified the contribution of both ports and airports in the status of global cities (Cartier 1999; Ducruet and Lee 2006; O'Connor 2010; Verhetsel and Sel 2009).

It is stated that gateways and hubs are emerging as a consequence of the changing role of places in the context of flows (Hesse 2010). Although gateway ports and transshipment hubs share functional characteristics, there are substantial differences between the two kind regarding the ability to develop logistics activities: while the literature confirms that gateway port-cities attract, which serve large hinterland markets, attract significant clusters of logistics firms, Slack and Gouvernal (2016) argue that (pure) transshipment hubs are rather fragile in attracting value-added logistics activities. They also underline some exceptions for their assumption: hubs such as Singapore, which is a major global logistics hub, thanks to its strategic location, modern infrastructures, financial incentives, free trade zones, etc.; a Singapore-hub-model copied by other transshipment hubs (see Chap. 6).

3.5   Global Trade and Port Network

The globalization trends and its world economy have led to a growing international trade and global market across the world, which extends the need for international transportation. Moreover, globalization has altered the traditional role of ports. Within the international trade route, maritime ports provide a low-cost and massive transport mean. Maritime transport continues to be an essential enabler of the global economy, since over 90% of the international trade is carried by seaports (International Maritime Organization 2012). However, the economic restructuring and technological development has also led to increase in cost of shipping. The traditional role of port, as the centre of trade activities, has been altered, due to the transportation revolution such as containerization and intermodalism. However, the port infrastructure remains an important mode of transport and logistics providing benefits at the local, regional and global level.

Though ports are developed within a city or city-region, an important factor in port-geography and port-economics relies on the fact that they belong to a port-system(s) and hierarchy, which reflects its functionality at the varied spatial scales. On the other hand, ports are physically dynamic infrastructures, and their spatial configurations have changed throughout the time. Moving beyond the traditional port models (see Chap. 2), UNCTAD (1992) and Van Klink (1998) have studied ports functional and spatial attributes while developing from a transport terminal to a logistics centre (hub) and identified four main phases: Table 3.2 summarises the key drivers, activities of the port, the dominant cargo flows and the spatial scale in

Table 3.2 Four stages in port's functional and spatial development

	Phase 1	Phase 2	Phase 3	Phase 4
Spatial scale	Port city	Port area	Port region	Port network
Period	Up to the mid nineteenth century	Mid nineteenth century to mid twentieth century	Late twentieth century	Late twentieth century, early twenty-first century
Key driver	Rise in trade	Industrialization	Globalization	Value-added Logistics
Main port functions	Cargo handling; storage; trade	Cargo handling; storage; trade; industrial manufacturing	Transhipment functions; cargo handling; storage; trade; industrial manufacturing; container distribution	Transhipment functions; Cargo handling; Storage; Trade; Industrial manufacturing; Container distribution; Logistics control
Principle cargo flow	General cargo	Bulk cargo	Containers	Containers and information

Source Adapted from Van Klink (1998: 147) and UNCTAD (1992)

each phase. The first and second phases are shaped simultaneously through trade and industrialization, as the spatial scale changes from a port city to port area with the expansion of port infrastructure and its land away from the historical city. In the third stage, the port regions are emerging as a response to the global forces. Mentioned beforehand, in port regions more functions are added to the port-area due to the increasing level of containerization. The fourth stage focuses on the new role of ports as a component in logistics and global supply chain (Hesse 2008).

In order to examine the contemporary port-system, here it is worthwhile making reference to studies that have made attempts in conceptualising transportation developments. The pioneering work by Taaffe et al. (1963), on the basis of the underdeveloped countries, applies their interpretation of an idea-typical sequence of transport development on some underdeveloped countries. Figure 3.3 illustrates the sequential phases in Taaffe six-stage model: (i) *scattered* settlements of small disconnected ports, emerged from the colonial occupation, with limited trading functions and restricted hinterlands; (ii) emergence of the first major *inland penetration lines* that leads to some ports and interior centres of population (I1 and I2) becoming more important than others, and the port concentration begins to develop (P1 and P2); (iii) *feeder routes* and *lateral connections* are evolving upon the major ports; economic development in the hinterland, along with the rising export trade; (iv) *rising interconnections* and development of intermediate centres into focal nodes (N1 and N2) for their own feeder networks; enlargement of major port hinterland; (v) a *complete interconnection* is formed as various feeder networks developing and linking around main ports and inland centres; (vi) national trunk-line routes or '*main streets*' are developed through a continuing lateral connection of all the major ports to the interior centres. Important centres would continue growing at the expense of other less competitive centres, and therefore 'high-priority' connections are emerging among major nodes.

The '*main street*' model conceptualizes the transport network development, highlighting the process of port concentration: from a group of small-disconnected ports along the coast, to the situation in which only a limited number (even one) of the major ports serve the extended hinterland, through an integrated transport network. With this model, Taaffe et al. (1963) explained how the port-system through inland connections determines port competition by the development of high-priority corridors between the biggest nodes. The hinterland expansion in the port regionalisation concept, reviewed in the previous chapter, is in fact a combination of load centres (Hayuth 1981) and high-priority corridors of the Taaffe et al. (1963) model, which is crucial in understanding the complex integration of nodes with different degrees of linkage in a transportation network, towards a hinterland control and competition.

In the current era of rising globalization and global economy, which has led to a substantial restructuring of transportation network along with a changing trend of the shipping line and their business pattern, a form of competition and co-operation is created among ports and in the port-related industry. Song (2003: 30–31) has outlined the driving forces behind this state of competition and co-operation: (i) *shipping alliances*; due to globalization and growing containerized trade, all ports are competing in one global market; large shipping companies are providing global

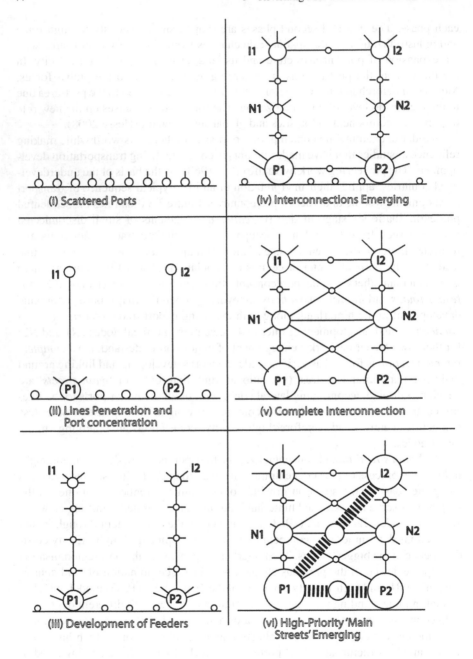

Fig. 3.3 An idea-typical model for the development of a transportation network in underdeveloped countries. *Source* Redrawn by the author from Taaffe et al. (1963: 504)

networks where one mega-carrier or an alliance can move goods freely around the global market; (ii) *larger vessels and intermodality*; fosters economies of scale; port's hinterland and foreland expands; increasing globalization of port management and operation; (iii) *intense port competition*; inter-related with previous factors; transformation of some feeder ports into regional hub ports or vice versa. Others have also stated that because of these forces, the shipping lines have been able to choose between ports competing to offer better situation: the biggest cranes, the deepest water and most favourable terminal leases (Hall and Robbins 2007).

3.6 Conclusion: Towards a Multidisciplinary Perspective on Port-City Studies

A large body of literature in urban studies has addressed globalization effects and the concept of world/global cities as commanding nodes in the world economy. However, the dynamic role of transport infrastructures in general, and ports in particular, in shaping the phenomenon of global cities is not clear. Although more than 90% of world trade volume is carried by the means of seaports (International Maritime Organization 2012), yet air transportation continues to become the preferred indicator in ranking global cities (Dogan 1988; Keeling 1995). On the other hand, the growing importance of seaports, as backbone of world trade, raises questions regarding their impact on the changing surrounding territory, as well as their dynamic role within the logistics and the global supply chain. Some scholars argue that the in a capitalist world economy, the rise in the volume of trade generates urbanization growth in dependent territories (Clark 2003).

As discussed in the previous chapters two and three, ports infrastructure and port-cities have been studied extensively within a wide span of disciplines. A great part of these studies deals with the waterfront redevelopment projects as the frontline of urban regeneration, fostered by the global economy and industrial shifts. On the one hand, urban planners and geographers have examined the new urban waterfront as gentrified spaces for more urban functions. On the other hand, the studies carried out by transport and economic geographers are centred on the changing position of ports within the global inter-port network structure (maritime networks, transport corridors, etc.). Port-centric studies, in general, deal with issues related to the new port infrastructure and regional development; port's socio/economic impact; port hierarchy and the inter-port network structure and etc. Different disciplinary studies tend to explore ports and cities as two separated assets. Indeed, ports are not only embedded within the global logistics, but also occupy urban spaces and contribute to local and regional economies. Thus, in order to understand the current and future challenges in port-city relationships, there's a need for a more integrated approach, beyond the merely waterfront or transport-economic studies.

References

Akhavan M (2017) Development dynamics of port-cities interface in the Arab Middle Eastern world—the case of Dubai global hub port-city. Cities 60(part A):343–352. https://doi.org/10.1016/j.cities.2016.10.009

Akhavan M, Mariotti I (2015) The role of infrastructural investment in attracting FDI: the case of Dubai. In: The XVII conference of the Italian association of transport economics and logistics (SIET). Bocconi University, Milan

Akhavan M, Ghiara H, Mariotti I, Sillig C (2020a) Logistics global network connectivity and its determinants. A European city network analysis. J Transp Geogr 82:1–9. https://doi.org/10.1016/j.jtrangeo.2019.102624

Akhavan M, Mariotti I, Ghiara H, Musso E, Silling C (2020b) Attractiveness of port-centric advanced logistics clusters. In: Wilmsmeier G, Monios J (eds) Geographies of maritime transport. Edward Elgar Publishing, Cheltenham and Northampton, pp 275–290

Alexander JW (1954) The basic-nonbasic concept of urban economic functions. Econ Geogr 30(3):246–261. https://doi.org/10.2307/141870

Amin A, Thrift N (2002) Cities: reimagining the urban. Polity Press, Cambridge

Andrews RB (1953) Mechanics of the urban economic base: historical development of the base concept. Land Econ 29(2):161–167. https://doi.org/10.2307/3144408

Antoine S, Sillig C, Ghiara H (2017) Advanced logistics in Italy: a city network analysis. Tijdschr Econ Soc Geogr 108(6):753–767. https://doi.org/10.1111/tesg.12215

Balasubramanyam VN, Salisu M, Sapsford D (1996) Foreign direct investment and growth in EP and is countries. Econ J 106(434):92–105

Barbieri E, Pollio C (2015) Industrial development and manufacturing in Chinese territories: the contribution of special economic enclaves policies. Working papers 1501, c.MET-05. Centro Interuniversitario di Economia Applicata alle Politiche per L'industria, lo Sviluppo locale e l'Internazionalizzazione, revised Jan 2015

Bassens D (2013) The city-upon-the-gulf: the relational growth and decline of 'world city' Dubai. In: Acuto M, Steele W (eds) Global city challenges: debating a concept, improving the practice. Palgrave Macmillan, Houndmills and New York, pp 47–62

Beaverstock JV, Smith RG, Taylor PJ (1999) A roster of world cities. Cities 16(6):445–458. https://doi.org/10.1016/S0264-2751(99)00042-6

Belloumi M (2014) The relationship between trade, FDI and economic growth in Tunisia: an application of the autoregressive distributed lag model. Econ Syst 38(2):269–287

Bichou K, Gray R (2004) A logistics and supply chain management approach to port performance measurement. Marit Policy Manag 31:47–67. https://doi.org/10.1080/0308883032000174454

Bird J (1963) The major seaports of the United Kingdom. Hutchinson, London

Burghardt AF (1971) A hypothesis about gateway cities. Ann Assoc Am Geogr 61(2):269–285. https://doi.org/10.1111/j.1467-8306.1971.tb00782.x

Carbone V, De Martino M (2003) The changing role of ports in supply-chain management: an empirical analysis. Marit Policy Manag 30:305–320. https://doi.org/10.1080/0308883032000145618

Cartier C (1999) Cosmopolitics and the maritime world city. Geogr Rev 89:278–289. https://doi.org/10.2307/216092

Castellani D, Zanfei A (2006) Multinational firms, innovation and productivity. Edward Elgar Publishing, Cheltenham

Castells M (1989) The informational city: information technology, economic restructuring, and the urban-regional process. Basil Blackwell, Oxford

Castells M (1996) The rise of the network society: the information age: economy, society, and culture, vol 1. Wiley, Chichester

Caves R (1996) Multinational enterprise and economic analysis. Cambridge University Press, Cambridge

Chen L, Notteboom T (2012) Determinants for assigning value-added logistics services to logistics centers within a supply chain configuration. J Int Logist Trade 10(1):3–41. https://doi.org/10. 24006/jilt.2012.10.1.001

Cho H, Ha Y (2009) Determinants of FDI inflow in regional port with resource-based view and institutional theory: a case of Pohang-Yeongil port. Asian J Ship Logist 25(2):305–331

Christaller W, Baskin CW (1966) Central places in southern Germany. Prentice Hall

Clark C (1945) The economic functions of a city in relation to its size. Econometrica 13(2):97–113. https://doi.org/10.2307/1907009

Clark D (2003) Urban world/global city. Routledge, London

Cooley CH (1894) The theory of transportation. American Economics Association

Council of Supply Chain Management Professionals (2007) Internet web site, supply chain management definitions

Coyle J, Novack R, Gibson B, Bardi E (1999) Transportation: a supply chain perspective. Cengage Learning

Cullinane K, Song D-W (2002) Port privatization policy and practice. Transp Rev 22(1):55–75. https://doi.org/10.1080/01441640110042138

Danielis R, Gregori T (2013) An input-output-based methodology to estimate the economic role of a port: the case of the port system of the Friuli Venezia Giulia Region, Italy. Marit Econ Logist 15(2):222–255

Dogan M (1988) Giant cities as maritime gateways. In: Dogan M, Kasarda JD (eds) The metropolis era: a world of giant cities. Sage, London, pp 30–55

Ducruet C, Lee S (2006) Frontline soldiers of globalisation: port–city evolution and regional competition. GeoJournal 67(2):107–122

Easterling K (2014) Extrastatecraft: the power of infrastructure space. Verso Books

Fageda X (2005) Load centres in the Mediterranean port range: Ports hub and ports gateway. Public Policies and Economic Regulation Research Unit, University of Barcelona

Fleming DK, Hayuth Y (1994) Spatial characteristics of transportation hubs: centrality and intermediacy. J Transp Geogr 2(1):3–18. https://doi.org/10.1016/0966-6923(94)90030-2

Freeman LC (1979) Centrality in social networks conceptual clarification. Soc Netw 1(3):215–239

Friedmann J (1986) The world city hypothesis. Dev Change 17:69–83. https://doi.org/10.1111/j. 1467-7660.1986.tb00231.x

GaWC (2010) Globalization and World Cities research group. Available at https://www.lboro.ac. uk/gawc/

Haezendonck E (2001) Essays on strategy analysis for seaports. Coronet Books Incorporated

Hall PG (1966) The world cities. McGraw-Hill

Hall PV, Hesse M (2013) Cities, regions and flows. Routledge, Oxford

Hall PV, Robbins G (2007) Which link, which chain? Inserting Durban into global automotive supply chains. In: Wang JJ (ed) Inserting port-cities in global supply chains. Ashgate, Hampshire

Hayuth Y (1981) Containerization and the load center concept. Econ Geogr 57(2):160–176

Heaver TD (2002) The evolving roles of shipping lines in international logistics. Int J Marit Econ 4(2002):210–230. https://doi.org/10.1057/palgrave.ijme.9100042

Heenan D (1977) Global cities of tomorrow. Int Execut 19(3):21–22. https://doi.org/10.1002/tie. 5060190313

Hesse M (2004) Land for logistics: locational dynamics, real estate markets and political regulation of regional distribution complexes. Tijdschr Econ Soc Geogr 95(2):162–173. https://doi.org/10. 1111/j.0040-747X.2004.t01-1-00298.x

Hesse M (2008) The city as a terminal: the urban context of logistics and freight transport. Ashgate, Aldershot

Hesse M (2010) Cities, material flows and the geography of spatial interaction: urban places in the system of chains. Glob Netw 10(2010):75–91. https://doi.org/10.1111/j.1471-0374.2010.002 75.x

Hesse M, Rodrigue J-P (2004) The transport geography of logistics and freight distribution. J Transp Geogr 12(3):171–184. https://doi.org/10.1016/j.jtrangeo.2003.12.004

Hesse M, Rodrigue J-P (2006) Global production networks and the role of logistics and transportation. Growth Change 37(4):499–509. https://doi.org/10.1111/j.1468-2257.2006.003 37.x

Holl A, Mariotti I (2017) The geography of logistics firm location: the role of accessibility. Netw Spat Econ 18(2):337–361

Hopkins AG (ed) (2003) Globalization in world history. W. W. Norton & Company, New York. ISBN 978-0-393-97942-8

Hoyle BS (1989) The port—city interface: trends, problems and examples. Geoforum 20(4):429–435. https://doi.org/10.1016/0016-7185(89)90026-2

Hymer S (1972) The multinational corporation and the law of uneven development. In: Bhagwati JN (ed) Economics and world order. Macmillan, New York, pp 113–140

International Maritime Organization (2012) International shipping facts and figures—information resources on trade, safety, security, environment. Maritime Knowledge Centre

Irwin MD, Hughes HL (1992) Centrality and the structure of urban interaction: measures, concepts, and applications. Soc Forces 71:17–51. https://doi.org/10.2307/2579964

Jacobs W, Hall PV (2007) What conditions supply chain strategies of ports? The case of Dubai. GeoJournal 68(4):327–342. https://doi.org/10.1007/s10708-007-9092-x

Jacobs W, Ducruet C, De Langen P (2010) Integrating world cities into production networks: the case of port cities. Glob Netw 10(1):92–113. https://doi.org/10.1111/j.1471-0374.2010.00276.x

Keeling DJ (1995) Transportation and the world city paradigm. In: Knox PL, Taylor PJ (eds) World cities in a world system. Cambridge University Press, Cambridge, pp 115–131

Ken WL (1978) Singapore: its growth as an entrepot port, 1819–1941. J Southeast Asian Stud 9(1):50–84. https://doi.org/10.2307/20070245

Keshavarzian A (2010) Geopolitics and the genealogy of free trade zones in the Persian Gulf. Geopolitics 15:263–289. https://doi.org/10.1080/14650040903486926

Knox PL, Taylor PJ (1995) World cities in a world-system. Cambridge University Press, Cambridge

Lavissière A, Rodrigue J-P (2017) Free ports: towards a network of trade gateways. J Ship Trade 2(1):7. https://doi.org/10.1186/s41072-017-0026-6

Lipsey RE (2000) Inward FDI and economic growth in developing countries. Transnatl Corp 9(1):67–95

Mangan J, Lalwani C, Fynes B (2008) Port-centric logistics. Int J Logist Manag 19:29–41. https://doi.org/10.1108/09574090810872587

Marcuse P, Van Kempen R (2000) Globalizing cities: a new spatial order. Wiley

Mariotti I (2015) Transport and logistics in a globalizing world: a focus on Italy. Springer International Publishing

Mariotti S, Piscitello L (2006) Multinazionali, innovazione e strategie per la competitività. Il Mulino

Muller EK (1976) Selective urban growth in the Middle Ohio Valley, 1800–1860. Geogr Rev 66(2):178–199. https://doi.org/10.2307/213579

Nam H-S, Song D-W (2011) Defining maritime logistics hub and its implication for container port. Marit Policy Manag 38(3):269–292. https://doi.org/10.1080/03088839.2011.572705

Notteboom T, Rodrigue J (2005) Port regionalization: towards a new phase in port development. Marit Policy Manag 32(3):297–313

Notteboom TE, Winkelmans W (2001) Structural changes in logistics: how will port authorities face the challenge? Marit Policy Manag 28(1):71–89. https://doi.org/10.1080/03088830119197

O'Connor K (2010) Global city regions and the location of logistics activity. J Transp Geogr 18(3):354–362. https://doi.org/10.1016/j.jtrangeo.2009.06.015

O'Connor K, Derudder B, Witlox F (2016) Logistics services: global functions and global cities. Growth Change 47(4). https://doi.org/10.1111/grow.12136

Olivier D, Slack B (2006) Rethinking the port. Environ Plan A 38(8):1409–1427. https://doi.org/10.1068/a37421

Panayides PM, Song D-W (2008) Evaluating the integration of seaport container terminals in supply chains. Int J Phys Distrib Logist Manag 38:562–584. https://doi.org/10.1108/096000308 10900969

Papadopoulos N (1987) The role of free zones in international strategy. Eur Manag J 5(2):112–120. https://doi.org/10.1016/S0263-2373(87)80074-9

Parnell S, Robinson J (2012) (Re)theorizing cities from the global south: looking beyond neoliberalism. Urban Geogr 33(4):593–617. https://doi.org/10.2747/0272-3638.33.4.593

Reed HC (1981) The preeminence of international financial centers. Praeger, New York

Rimmer P (1999) The Asia-Pacific Rim's transport and telecommunications systems: spatial structure and corporate control since the mid-1980s. GeoJournal 48:43–65

Robinson R (2002) Ports as elements in value-driven chain systems: the new paradigm. Marit Policy Manag 29(3):241–255. https://doi.org/10.1080/03088830210132623

Sanchez RJ, Hoffmann J, Micco A, Pizzolitto GV, Sgut M, Wilmsmeier G (2003) Port efficiency and international trade: port efficiency as a determinant of maritime transport costs. Marit Econ Logist 5(2):199–218

Sassen S (1991) The global city: New York, London, Tokyo. Princeton University Press, New Jersey

Sassen S (1994) Cities in a world economy. Pine Forge Press, Thousand Oaks

Scott AJ, Storper M (2014) The nature of cities: the scope and limits of urban theory. Int J Urban Reg Res. https://doi.org/10.1111/1468-2427.12134

Shin K-H, Timberlake M (2000) World cities in Asia: cliques, centrality and connectedness. Urban Stud 37(12):2257–2285. https://doi.org/10.1080/00420980020002805

Short JR, Breitbach C, Buckman S, Essex J (2000) From world cities to gateway cities: extending the boundaries of globalization theory. City 4(3):317–340. https://doi.org/10.1080/713657031

Sigler TJ (2013) Relational cities: Doha, Panama City, and Dubai as 21st century entrepôts. Urban Geogr 34(5):612–633. https://doi.org/10.1080/02723638.2013.778572

Slack B, Gouvernal E (2016) Container transshipment and logistics in the context of urban economic development. Growth Change 47(3):406–415. https://doi.org/10.1111/grow.12132

Smith J (1910) The world entrepôt. J Polit Econ 18(9):697–713

Smith MP (2001) Transnational urbanism: locating globalization. Blackwell Publishers, p 221

Song D-W (2003) Port co-opetition in concept and practice. Marit Policy Manag 30(1):29–44. https://doi.org/10.1080/0308883032000051612

Taaffe EJ, Morrill RL, Gould PR (1963) Transport expansion in underdeveloped countries: a comparative analysis. Geogr Rev 53(4):503–529

Tan T-Y (2007) Port cities and hinterlands: a comparative study of Singapore and Calcutta. Polit Geogr 26(7):851–865. https://doi.org/10.1016/j.polgeo.2007.06.008

Taylor PJ (2004) World city network: a global urban analysis. Routledge

von Thunen JH (1826) Der Isolierte Staat in Beziehung auf Landwirtschaft und Nationalökonomie, Schumacher-Zarchlin H (1875), Wiegandt, Hempel und Parey. Williamson OE (1985) The Economics of Institutions of Capitalism, the Free Press, New York

Tsang S (2007) A modern history of Hong Kong. I. B. Tauris

UNCTAD (1992) Port marketing and the challenge of the third generation port. In: United Nations conference on trade and development, committee on shipping, New York

UNCTAD Port Section (1996) UNCTAD monographs on port management

United Nations (2005) Free trade zone and port hinterland development. United Nations ESCAP. https://books.google.nl/books?id=waEUu-A6RfwC

Van Klink HA (1998) The port network as a new stage in port development: the case of Rotterdam. Environ Plan A Econ Space 30(1):143–160. https://doi.org/10.1068/a300143

Verhetsel A, Sel S (2009) World maritime cities: from which cities do container shipping companies make decisions? Transp Policy 16(5):240–250. https://doi.org/10.1016/j.tranpol.2009.08.002

Wang JJ, Cheng MC (2010) From a hub port city to a global supply chain management center: a case study of Hong Kong. J Transp Geogr 18(1):104–115. https://doi.org/10.1016/j.jtrangeo.2009.02.009

World Bank (2008) Special economic zones: performance, lessons learned, and implications for zone development. The World Bank, Washington

Yeoh BSA (2003) Contesting space in colonial Singapore: power relations and the urban built environment. Singapore University Press

Chapter 4
Making of a Global Port-City in the Middle East: The Dubai Model

4.1 Dubai Model: A Laboratory for Port-City Development Studies

Since its founding in the early nineteenth century, Dubai has developed with an unprecedented pace to emerge as the main trading hub in Middle Eastern region, as well as a major global player. Strategically located along the south eastern coast of the Persian Gulf—200 km south west of the Strait of Hormoz[1]—between Abu Dhabi and Sharjah (Fig. 4.1), Dubai with a current land area of approximately 4000 km^2 and over 2.8 million population (year 2019) is the second largest city estate among the seven emirates of the United Arab Emirates (UAE). Since 1950, Dubai population grew about 100 times from a small town of 20 thousand inhabitants to about 2 million inhabitants by 2010, while its urban built space extended rapidly 400 times approximately (Government of Dubai n.d.).

Starting from its formal creation in 1971, and thanks to the wealth brought by the oil income, United Arab Emirates has been growing with an unprecedented speed, to mark itself as one of the fastest developing nations, while currently it accounts for the second largest economy in the Arab World (after Saudi Arabia). The oil revenues heavily, invested in the so-called branded mega-projects (Ponzini 2011), has led to particular urbanization patterns, especially in Dubai and Abu Dhabi. As outlined in the introduction of this book, UAE and its major emirates, mainly Dubai and its rapid urbanization and construction boom has drawn great attention from researchers in varied disciplines, especially with respect to other Arab States of the Middle East; in this regards, architects, urban planners and designers have shown particular interest in the development strategies applied by the emirates, to boost business, real estate, tourism and trade (among others Acuto 2010, 2014; Bagaeen 2007). Though some scholars have made attempts to study the importance of trade

---

[1] Strait of Hormoz is the only sea passage from Persian Gulf to the open ocean and a vital transit point for world crude oil. With around 35% of the petroleum traded by sea passing through this straight, it is considered as an important strategic location for international trade.

Fig. 4.1 Satellite image of the Persian Gulf. *Source* Elaborated by the author, base image from Google Earth

and logistics in Dubai's making (Fernandes and Rodrigues 2009; Keshavarzian 2010; Thorpe and Mitra 2011; Ziadah 2018), only recently a very limited number of studies have actually considered Dubai as a 'port-city' (Akhavan 2017; Broeze 1999; Jacobs and Hall 2007; Ramos 2010; Sigler 2013) and yet still we know little about the interaction between its maritime ports and the city itself.

In this chapter, I shall discuss a Dubai model not with reference to its overall ambitious and accomplishments of the emirates; here, the focus is on the maritime port infrastructures, its particular port and city evolution and linkage, as well as the historical and key importance of its maritime ports in constructing a desert. Although the history of Dubai can be traced back to the 1800s, what concerns this study is the historical role of Dubai as a free port, since the 1900s. The entrepôt characteristics set the basis for developing an economy based on international trade, initiated with expansion of the Creek, followed by port constructions and expanding re-export activities through free trade zones. Here, the argument is centred on Dubai port-city evolution thorough a synergistic relationship between global and local forces within a particular historical and geo-political context. The four-stage port-city evolution introduced by Akhavan (2017) is, therefore, used as a tool to investigate and discuss the factors that characterise Dubai's port and city at each phase, as well as the changing spatial and functional dynamics at the port-city interface:

- Phase I (1900s–1950s): The freeport-merchants town
- Phase II (1960s–1970s): The entrepôt port-city
- Phase III (1980s–1990s): The transshipment hub port city
- Phase IV (2000s–present): The logistics hub port city.

4.2 Phase I (1900s–1950s): The Freeport-Merchants Town

Although the Middle East in general and the Persian Gulf region in particular has a rich history of urbanisation, there is not so much to say about Dubai port-city development prior to the oil discovery in the mid-1960s. Officially formed in 1971, UAE is yet still a young county; however, the history of this nation is believed to be quite rich (Peck 1986; King 1997). It is not the scope of this chapter to trace back the regional history, but it is important to highlight some historical attributes and turning points that set the foundation for Dubai's modern era.

This pre-oil phase is also marked by the colonial era, with the entrance of the Western power starting from the sixteenth century. More precisely, the European powers, starting with the Venetians, and subsequently the Portuguese, the Dutch, and finally the British Empire were interested in this region as a means to secure trade routes to and from the Indian and Eastern regions. In the period between 1763 and 1971, the British Empire became the dominant power, by maintaining varying degrees of political control over some of the states, including UAE, Bahrain, Kuwait, Oman and Qatar through the British Residency of the Persian Gulf (Hawley 1970). British interest in the region, primarily entered as commercial pursuit, increased mainly along the East Indian Company trade routes. With the increasing presence of France, Germany and Russia, in 1892, Britain and the sheikhs of the current UAE (known as the Trucial coast) signed a treaty, known as the 'Exclusive Agreement'. This agreement prohibited the UAE (Trucial) rulers from the right to cede territorial sovereignty without British consent. Furthermore, from this period until the independency in 1971, Britain assumed the responsibility for foreign relations and protection of the individual coastal sheikhdoms. Under this agreement, British's growing influence shifted from commercial to strategic priorities, which formed the diplomatic pillar of British power in the UAE region (Peck 1986: 36).

Historically in this region, the Persian ports were of great importance: Lingah port-city was the main trading hub between the East and West and the dominating centre for pearling trade (Davidson 2005). However, during the 1870s, and due to the needs for financial resources and the political instability in Iran, other ports along the Arab coast of the Gulf got the opportunity to become business and trading centres. As a consequence of the raised taxes in 1902, the Persian-Arab merchants—most importantly the Persian Bastak—immigrated to Dubai to establish a business base. In fact, the ruler of Dubai took advantage of the raised taxes and the merchants searching for another port in proximity, and therefore declared Dubai a *freeport* through applying a free trade policy—an exempt of all import and export tariffs— with the ambition to create a business-friendly port. Persian merchants play a key role in Dubai's prosperity, by consolidating the business environment and trading links; only few years after the advent of freeport in 1906, more than 7000 were involved in the pearling trade (Wilson 1999): this is also why Dubai is known as the 'City of Merchants'.[2] Thanks to such effective strategies, and also for other guaranteed benefits (such as free land), by 1925, the temporary settlement of the merchants became permanent; and later on during the 1930s, another influx of immigrants from the Indian sub-continent (the Baluchis) arrived and start working as porters and in other low-level occupation (Davidson 2005). In fact, becoming a freeport had made Dubai even more competitive with respect to other neighbouring emirates, Sharjah and Ras Al Khaimah, as they offered better ports at that time. Thus, after a short period, the number of ships that called at Dubai port raised from five times a year to twice a month (Krane 2009).

As the city's historical port continue to attract business and trade importance, there was an urgent need to widen and deepen the Dubai Creek to handle large vessels and improve the loading and unloading facilities. A plan to develop the Creek was, therefore conceived in 1955 to ensure dredging shallow areas, constructing breakwater and quay suitable for transshipment activities. Accordingly, the first project was the deepening of the Creek and removal of sand and silt deposits by an Austrian company, was completed in 1956 and worth 400 thousand pounds. The significant rise in the volume of cargo was a proof to the success of the project and therefore to confront this increasing demands for trade, the banks of the Creek were re-constructed for shops, warehouse and other port-related facilities; the revenues gained from the land sales were then used to pay back project loan (Doxiadis Associates 1985).

The pace of urbanization is relatively slow in this phase, during the years from the 1900s to 1950s. Nevertheless, the turn of the twentieth century is pinpointed by the birth of a free port, which is a crucial point for an increasing commercial activity, trade and therefore making the modern Dubai. From a spatial point of view, the city goes through a rather steady and natural development with limited urban expansion also because of the constrained economic growth, mainly based on fishing and pearl industry. The port-city interface is characterised by expansion around the Creek, which divides the city in two main agglomerations: the mercantile sector on the north (Deira) and the government centre on the south (Bar Dubai) of the Creek—this division is clearly visible in Dubai map of the 1950 (see Fig. 4.2), which becomes more realistic in Fig. 4.3 that shows the aerial view of the same year. Being attracted to preferred economic opportunities, during the 1900s, Dubai experiences the first major influx of immigrants (Pacione 2005); therefore, the first phase of port-city development: the small fishing town of 10,000 population of the early 1900s that spreads over only 0.2 km² of land, expanded over 3.2 km² with a population of around 50,000 by mid-1950s (Dubai Municipality; Doxiadis Associates 1985).

---

[2]https://www.dpworld.com/who-we-are/about-us.

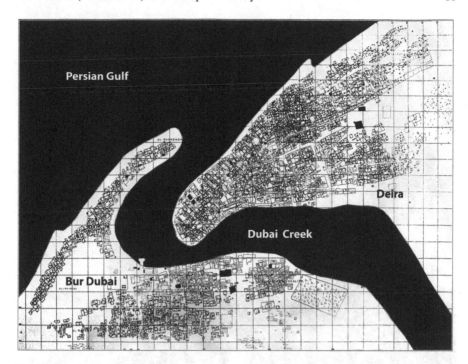

Fig. 4.2 The map of Dubai in 1950—the urban settlements around the Creek. *Source* Elaborated by the author, base map retrieved from: https://commons.wikimedia.org/wiki/File:Exhibition_(His torical_documents_center)_(Building_No_128)_-_picture_of_old_Dubai_map.jpg#filehistory

4.3   Phase II (1960s–1970s): The Entrepôt Port-City

The discovery of petroleum in the early 1960s, in coastal waters of Abu Dhabi, was followed by significant increase in commercial activities, determining the power structure and providing significant wealth to this sheikhdom, which still remains the largest emirate and capital of UAE. The oil in Dubai was founded in 1966, which brought an unprecedented wealth to the city. It was around this time that Britain failed in competing with U.S oil companies in related investments and contracts. This led to the weakening British power in the region, particularly the Arab emirates; by the early 1970s, Britain declared to terminate all the treaty agreements with the seven current emirates and to withdraw the British military forces from the region (Gornall 2011). Consequently, during 1971–1972, the seven Trucial states (Abu Dhabi, Ajman, Al Fujayrah, Dubai, Sharjah, Umm al Qaywayn and Ras al Khaymah) agreed on a federal constitution for achieving independence as the United Arab Emirates-UAE (Country Profile 2007). The space freed up by the removal of British military was immediately filled by United States and Russia. However, it is evident that the Persian Gulf's vast oil reserves make the area a continuing source of international tension.

This phase, starting from the 1960s, is marked by a rapid urbanization, substantial restructuring of the spatial structure as well as the socio-economic and political

Fig. 4.3 The aerial image of Dubai historic centre, taken around 1950. *Source* Image credit supplied by Gordon Brent Brochu-Ingram

dimensions. The surge of oil revenues coupled with the formation of a new nation (the UAE) set the basis for an oil-driven-urbanization, capital intensive infrastructure construction (most importantly the two modern ports) and city development through formal planning approaches. As seen in the first phase, the expansion of the Creek helped Dubai stabilising its position as one of the region's main port of call for bigger ships: increase in commodity flows and transshipment activities gave path to create an entrepôt port-city. Although the population had already exceeded 50,000 by the end of 1950s (Dubai Statistics Center), and the trade and commercial activities continue to increase, the city lacked basic infrastructures and urban facilities and yet still did not have a systematic urban plan. The British architect John Harris was, hence, appointed to design a master plan. Presented in 1960, the First Master Plan of Dubai (see Fig. 4.4) introduced a preliminary system of urban infrastructure, based on grid and zoning to ensure fundamental elements: pedestrian pathways, a road system, residential zones, commercial districts, industrial areas and a new town centre (Harris

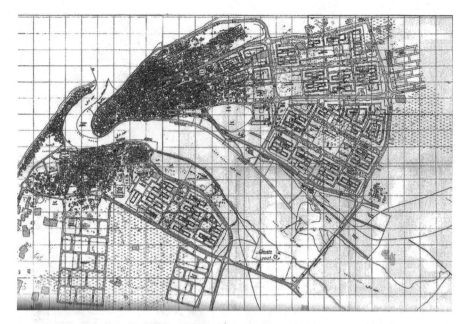

Fig. 4.4 John Harris's First Master Plan for Dubai (year 1960)

1994). In this plan, the Creek has a key role, being the centre of trade and economic livelihood of the port-city interface. The new designed grid, that follows the city's original palimpsest around the Creek, extends further towards the desert with more rigid urban block.

With the growing urbanization and oil discovery in 1966, the first Master Plan was seen incapable to guide such expansion and meet the desire of the Sheikh for a modern Dubai. The revised and upgraded version was then introduced in 1971: the Dubai Development Plan Review, which plays a key role in making the modern Dubai. As seen in Fig. 4.5, although the second Master Plan was mainly concerned about the area surrounding the Creek, it also foresees development along the coast and inland areas. This plan pays particular attention to transport infrastructure, in addition to the first plan, it includes: a tunnel, two bridges across the Dubai Creek (to connect the two major historic districts of Deira and Bur Dubai) and the first man-made modern port (Port Rashid). The plan also addresses the housing demands of the rising population (the colour blue in the plan indicates the housing land use): a large residential zone was envisioned—now known as Jumeirah—along with facilities for health, education, leisure and recreational activities (Gabriel 1987). Of particular importance was the construction of Dubai-Abu Dhabi highway (1971–1980); currently known as the Sheikh Zayed Road, it runs parallel to the coastline and stretches from the Trade Center Roundabout towards Abu Dhabi, 55 km away from the Jebel Ali area.

The construction of the first man-made mega port infrastructure—Port Rashid completed in 1972, with 12 general cargo berths which was then extended to 24 berths by the end of 70s—marks an important turning point in developing a more complex

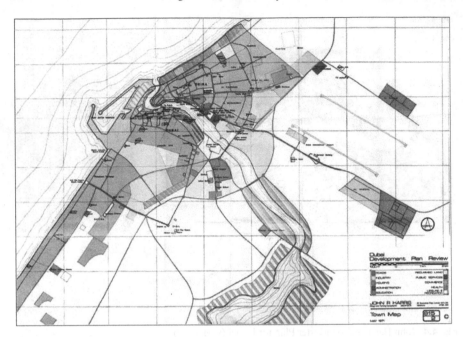

Fig. 4.5 John Harris's Second Master Plan for Dubai (year 1971)

port-city interface in Dubai's historic centre: the traditional cargo handling was upgraded to meet the requirements of advanced shipping industries and containerisation. The port was extensively used for importing large amount of construction materials (e.g. cement) for the real estate boom, giving the new port a key role in Dubai's modern development (Akhavan 2017).

The first attempts to construct port ancillary infrastructures in the port hinterland was seen in this phase: located in the southern wing of Port Rashid, the Dubai Dry Dock and Ship Repair facility was realized in 1979 as a joint venture with British companies. Developed at the mouth of the Creek, in the historic centre and the most populated part of the city, Port Rashid and the dry dock facilities, were part of the expansion projects after the Harris's second master plan (Ramos 2010). It is worth mentioning that besides Port Rashid and Dry Dock, the construction of the 'World Trade Center' (during the ate 1970s) represented a final contribute to Dubai's trilogy projects towards a modern city that was open to global business. Though the first location for this 40-storey tower was planned in proximity to the port at the mouth of the creek, but to overcome the height and space restrictions it was finally placed slightly beyond the city limits. As the first tower-landmark, the World Trade Center became the first break in the Harris's plan, giving path to a new southwestern growth axis.

The concentration of new infrastructures within the historic centre, on the one hand, led to a more sophisticated trade and logistics; on the other hand, the increasing commodity flows generated negative effects, such as congestion and pollution. This was partly solved by the construction of the new maritime port far from the city centre. The project for Mina Jebel Ali was therefore launched in 1976, destined to be the world's largest man-made port, with the aim to increase trade and throughput capacity, develop an industrial complex and continue the urban expansion towards the western border. Opened in 1979, with 15 km of waterfront and 67 berths, the Jebel Ali Port is located 35 km far from the city centre, near the city's border with Abu Dhabi (Broeze 1999). The original plan, as sketched in Fig. 4.6, was a new town—including a seaport, airport, industrial and residential areas—to bring industrial development, while Dubai remains the social, commerce and administrative centre (Ramos 2010). Although the new town project was never realised, the port and industrial zones were constructed following Dubai's free port strategy. Furthermore, a residential complex, known as Jebel Ali village, was also built (in 1977) to house expatriates and the construction workers.

This phase—during the 1960s and 70s—is characterised by a rapid urbanization and the capital-intensive infrastructural development. The construction of two new ports on the either end of the city laid the foundation for developing a modern desert along the coast. From an urban planning perspective, in these two decades the city experiences its first formal urban plans—structural planning and master plans—influenced by Western architects and planners. This also reflects the power of local governance that is open to and relies international thinking when it comes to systematic urban growth. From the systematic planning here emerges two rather different port-city interface: (i) the historic centre (the Creek + Port Rashid + Dry Dock), which is defined by the modernized port and ancillary infrastructures in a dense urban settlement; (ii) the Jebel Ali suburban area, where a whole new trade-based complex—that consists a massive port, industrial areas and residential districts—is

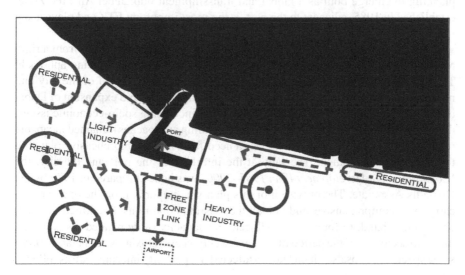

Fig. 4.6 The concept of Jebel Ali new town. *Source* Redraw by the author—original image from Ramos (2010: 120)

built from scratch to become a growth pole. The first major highway—Sheikh Zayed Road—that runs through the city, represents an important element growth corridor to attract commercial and financial activities (Alshafieei 1997). Thanks to these top-down approaches and rapid constructions, by the late 1970s, the city expanded to cover 84 km^2 and the population had grown to about quarter of a million (Doxiadis Associates 1985). The urbanization in this phase is also marked by the second phase of major influx of immigrants moving to Dubai for mainly as workers in low-skilled occupations, from India, Pakistan and other Arab states were imported to meet the needs of the booming construction sector (Pacione 2005).

4.4 Phase III (1980s–1990s): The Global Transshipment Hub Port City

Already from the early 1980s, the twin mega-ports of Dubai, situated 35 km from one another, started acting as complementing trade-infrastructures, and show success in attracting container traffic: the total throughputs grew from 4500 TEUs (in 1976) to around 64,000 TEUs (in 1980), with 70% of transshipment activities (Cuthbert 2011). On the other hand, in this phase, Dubai became a major entrepôt of the Persian Gulf region (see Sect. 4.5), as it handled 39% of the regional traffic (throughput) by the mid-90s (Akhavan 2017). In line with Dubai's historical freeport policies, of particular note in this phase is the proliferation of free trade zones (hereinafter FTZ) to attract foreign direct investment (hereinafter FDI) and meet the needs of the growing trade; a significant stage for a post-oil development in a city, which is planning to emerge both as a global and transshipment hub: Jebel Ali Free Zone, established in 1985, adjacent to the port, is the most important FTZ and plays a key role in today's logistics sector.

From an urban planning perspective, this phase is marked by a shift from a rigid Master Planning to a more comprehensive and urban structure planning approach. The previous plans were mainly focused on the existing situation of the city and neglected the regional and future growth. Accordingly, the rapid expansion of Dubai beyond the borders of the rigid master plans and increasing socio-economic issues, due to the vast influx of immigrants, the past planning approaches remained incapable of managing the city growth. There was a need for a more comprehensive planning to guide the development, specially at the interface of the port and the city: the *Comprehensive Development Plan 1985–2000*' was hence prepared in 1985 by the Doxiadis Associate. The provision of this plan was important, on the one hand, to carry out a comprehensive study on all the existing conditions and difficulties and, on the other hand, to forecast and allocate the capacity of demanded resources by different sectors over the period of the next twenty years (Doxiadis Associates 1985); some specific proposals, which later on affected the port-city interface, are as follows:

- Residential projects for foreign labours, which are living in temporary labour camps with minor living qualities.

- Cleaning the water of Jumeirah Beach, polluted from the port-generated waste, to become a pilot project for recreation facilities in the city.
- Removing the industrial area from the congested part of the city and developing new industrial zones in parts away from the city centre, such as Jebel Ali area.

Based on this comprehensive plan, in the early 1990s, the government commissioned the Structure Plan for *Dubai Urban Area 1993–2012* to the consultants Parsons Harland Bartholomew & Associates, with the aim to guide the economic and spatial development of Dubai into a twenty-first century post-oil city; with construction of mega projects as development tool to create 'cities within cities'. As seen in Fig. 4.7, this urban structure plan is based on zoning, with a series of nodes and axes for growth, which sets the foundation of the city pattern of what appears today (Elsheshtawy 2010). The plan, therefore, provided a land-use structure for the city, based on some key challenges concerning the spatial growth: (i) allocating additional land to accommodate urban expansion, by considering the current and future demands (land-uses to year 2012 and beyond) for residential, industrial and commercial functions; (ii) expanding the current transport network and infrastructure facilities, and those concerning the economic growth; (iii) boosting a consistent and sustainable economic growth; (iv) supporting and attracting the private sector to participate in the process of urban development; (v) promoting public–private cooperation in the construction of megaproject; (vi) encouraging foreign investments in local enterprises, also from an institutional point of view; (vii) creating an inter-departmental organisation for reviewing, monitoring and implementing the Urban Structure Plan; (viii) establishing a regulatory framework considering the needs of varied agencies and organizations (Pacione 2005).

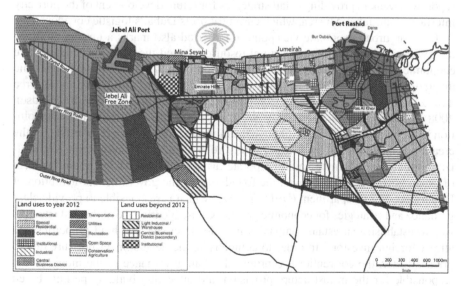

Fig. 4.7 Dubai urban structure plan 1993–2012 (the two port-city interfaces are highlighted). *Source* Elaborated by the author, original image from Pacione (2005: 263)

Once again, as highlighted in Fig. 4.6, from this plan, the two different port-city interfaces—connected through the two main corridors of Sheikh Zayed Road and the Inner Ring Road—are apparent:

- *The Port Rashid Interface*: it has historically been the main commercial centre of the city. The 2012 land use plan also confirms this, as well as the concentration of residential and recreational activities, and also a major business, banking and administrative centre (developed in the historic district of Deira). The city's first airport is almost located in this area (with a limited industrial area adjacent to it). Beyond 2012, the plan foresees secondary central business districts and multi-functional complexes, mainly for the business and tourism industry.
- *The Port Jebel Ali Interface*: it was originally born an industrial district and the 2012 land use plan is a proof; Dubai's second major airport is located in this zone. The co-location and concentration of the industries (within free trade zones) in close proximity to the two main maritime port and airport ensures reduction in costs and economies of scale. The structure plan proposes residential areas, beyond 2012, to be located in the far western part with a linear public open space (corridor) that acts as a buffer, with limited tourism facilities.

The report on the plan foresees a spatial growth development along the coast (now the New Dubai), and beyond the Jebel Ali area. In brief, it proposes three main nodal development centres: east of Mina Seyahi (the current Pam Jumeirah), area of Ras Al Khor and south of Jebel Ali. The first two areas were planned as multi-functional zones, a combination of public spaces, office, cultural, and other recreational facilities. As noted earlier, for the Jebel Ali area, large areas are dedicated to residential functions, the airport complex (now the Dubai World Central); nevertheless, the report was weak in providing detail strategies for future development of the port-city interface at the Jebel Ali area, which currently hosts Dubai's logistic corridor.

From an urban economic viewpoint, this period also marks a turning point, as the 1980s revelation of limited natural resources urged the need for an alternative economic structure replacing the revenues from the oil. The strategies for economic diversification was foreseen to enhance trade and business, by developing various free trade zones, tourism, and megaprojects through freehold real estate market (Davidson 2008). From an institutional perspective, in this period the city experiences a revolution in the urban governance. Although the municipality has been responsible for the creation of the plans, yet by the early 90s there was a transformation, from a centralized administrative form to a more corporate-driven approach. The restructuring form of urban governance started with the foundation of the governmental institution of Dubai Economic Department (DED) in 1992, which is responsible to provide clear guidance and strategies for economic growth. Consequently, one of the first attempts was to establish real estate companies in order to create free-hold properties and attract foreign investors. In order to compensate the development pressure on the private sector, the conventional centralized urban governance of the municipality, responsible for the infrastructure provision and allocating building permits based on the Structure Plan of 1993, had to be restructured to a more decentralized and flexible urban management. The result was the creation of large holding real-estate companies,[3] which have played a key role in urban growth through megaprojects.

Thanks to the wealth brought by the oil, Dubai saw a construction boom in this phase which, on the other hand, grew even more the influx of migrant—mainly low-skill and low-paid—workers, to meet the high demands of the construction sector. Therefore, by the turn of the century, the population had increased to about 800,000—more than 50% expatriates—with an urban area that extended over 178 km^2 (year 1993). The main driving forces for the urban development in this phase was the Jebel Ali Free Zone and the Emirates airline, both established in 1985. By 2000, Dubai becomes the main transport hub (both freight and passenger): the Jebel Ali Port increases its capacity to 3 million TEUs; the airport has also expanded to handle more than 12 million passengers. Here, it is worth mentioning that, from the end of this stage onward, Western professionals and high-skilled workers were attracted as floating citizens to help developing the economic diversification.

4.5   Phase IV (2000–Present): The Logistics Hub Port City

As stated beforehand, in the second phase, the oil revenues made possible the construction of maritime ports and its ancillary infrastructures to support the trade and set the foundation for an economic prosperity. At the beginning of this phase Dubai was hit by the global economic recession and burdened major negative impacts concerning its growth path strategy (Faris and Soto 2016), yet still the 2000s is Dubai's golden urban development both spatially as well as the diversity of megaprojects planned and realized; it is at the beginning of this stage that Dubai becomes exceeds one million population (in 2001). The city therefore marks itself as the first post-oil economy in the Middle East (Krane 2009), as a result of the strategies to diversify its economy: oil stands for only 10% of the current GDP (Dubai statistics yearbook 2013), while shifting from oil-dependent to a new economy based on trade, business and tourism. The expansion of FTZs continues in this phase, mainly in close proximity to the maritime ports and airports, which marks an important factor while studying the evolution of port-city interface.

Such development strategy necessitated private investments. The continuing privatization of the urban governance, which had started in the 1990s has led to a growing decentralized pattern of spatial and economic development. From an urban planning point of view, this period is characterized by the need for a more strategic planning, since the previous structural planning proved deficiencies in coping with the rapid urbanization and economic development in Dubai. Henceforth, in 2007, the executive council commissioned the Dubai Strategic Plan 2015 (hereinafter DSP),

---

[3]The large holding real estate companies in Dubai falls under the government umbrella and are typically public joint stock companies. The three most important are Nakheel, Emaar Properties and Dubai Holding, which were founded in the 90s under the control of the ruling family.

as the first long-term strategy, which best outlines the city ambitious along five main pillars of strategic priorities (Government of Dubai n.d.):

1. *Economic development*: GDP growth; productivity improvements; economic stability; competitiveness
2. *Social development*: national identity and social cohesion; Emiratis role in the society and economy; availability of quality education, health and social services; work environment; cultural life
3. *Infrastructure, land and environment*: urban planning; energy and water; transportation; environment and waste management
4. *Security, justice and safety*: security and order; protection of civil rights and freedom; crisis and disaster management; equality and expedition in judicial system; public health; safety
5. *Governance excellence*: strength strategic and forward-looking focus; enhance organizational structure and accountability; increase efficiency; enhance responsiveness and customer service; empower and motivate public service employees.

In preparing the DSP, the economic guidelines were designed regarding some specific factors such as: Dubai's GDP growth between 2000 and 2005; professional human resource, mainly driven by the government policies to foster investment and business; as well as the construction of special economic zone and mega projects. There is an emphasis on reducing the dependence on oil and deliberate policy in diversifying the economy, along with sector specific development strategies. As described by the planning report, the growth strategies envision Dubai as a regional business hub, an attractive tourism destination and safe place to invest and live. Hence, the transportation sector is given a strategic vision 'to integrate infrastructure development and environmental focus in order to achieve sustainable development'. The guidelines provided by the plan, generally, address the tourism sector, real estate and housing provision. Although the logistics sector is the key economic force, however the report fails in defining strategies for freight flows and related infrastructure provision. The DSP also considers the private sector participation as the secondary engine in boosting the economy; however, due to the active presence of the government, role of the private institutions in different sectors are not clear.

The most recent urban planning tool is the *Dubai Urban Development Master Plan 2020*, approved by the Dubai Executive Council in 2011; the 'Supreme Urban Planning Council' was therefore formed to streamline the urban and environmental planning process. Based on the report, Dubai Emirates main land covers 3978 km^2, of which 20% approximately covered by the existing urban built space and projects under construction. The undeveloped land (which was committed before 2008 for mega projects) covers also 20% approximately of the Emirate main land area (Government of Dubai n.d.). Highlighted in Fig. 4.8, the plan distinguishes four main areas, based on their land-use and strategies for future development:

- *Area 1—Offshore Islands*: this is the sea territory subject to dredging and reclamation, and houses Dubai's artificial islands (known as the Palms Islands), which expands the waterfront of the city; mainly for residential, resorts and tourism.

Fig. 4.8 Urbanization parameters for Dubai urban spatial structure plan 2020 and beyond. *Source* Elaborated by the author, original image from Government of Dubai (n.d.). Dubai 2020 urban master plan

- *Area 2—Metropolitan Area*: the linear city expands along the waterfront and covers the existing urban fabric, on-going and on-hold megaproject.
- *Area 3—Non-Urban Area*: this pre-urban area in the desert is characterised by the landscape of traditional sport facilities, resorts, conservation areas, non-urban settlements and other special activities, etc.
- *Area 4—Non-Urban Area Desert*: this non-urbanized area includes conservation areas, resorts, gas extraction area, aquifer zone, farming settlements, etc.

The three key drivers for the 2020 plan are: (i) a vision for Dubai *as a modern Arab city and vibrant regional gateway to the world*; (ii) an integrated city and regional development planning framework; (iii) and a legal and institutional framework. The plan seeks to improve social, economic and environmental sustainability by directly addressing transportation, housing affordability, culture integration and waste management. This is the first plan for Dubai that addresses any of these key issues. The Plan considered the impact of the global economic downturn in 2008; with the aim to promote flexible and responsive land-use, and quality-built environment (Government of Dubai n.d.). However, there is limited information available about the 2020 plan and its secrecy is unsettling, especially in contrast to The Abu Dhabi Plan 2030 that has been made available to the general public for review and comment.

In this final stage of Dubai port-city development, the expansion of the Jebel Ali area into an intermodal infrastructural basis is of great importance. In fact, this area has been one of the main zones of development, also proposed by the Structure

Plan for Dubai Urban Area 1993–2012. However, the most recent intervention is the multimodal integrated logistics platform—Dubai Logistics Corridor (see Chap. 5)— inaugurated in 2010, with the vision to host around 10,000 companies with 305,000 employees, is an attempt to shift from a transshipment hub towards a more specialized logistics hub port within the region. With the aim to link sea, land and air, this hinterland includes:

- *Port Jebel Ali* is the massive man-made freeport infrastructure that is the main (hub) container terminal of the Region.
- The two free trade zone areas—*Jebel Ali Free Zone (JAFZ)* North and South, adjacent to the Jebel Ali Industrial Area which are of the first attempts to develop the port hinterland and logistics activities in Dubai; main functions of JAFZ include port discharge cargo, warehousing and logistics, trade, manufacturing and service for the local, regional and international players.
- The new airport complex *Dubai World Central* (hereinafter DWC); launched in 2006, at the centre of this 140 km² multifunctional development is located *the Al Maktoum Airport* (officially opened in 2010 for freight flows, followed by passenger flights in 2013), which is bounded by the aviation, logistics and residential districts. Currently known as the 'Dubai South', it is established as a free zone for business and organizations operating in the aviation and logistics industries. The master plan, now refined, originally had a working title of Jebel Ali Airport City. It is designed to support Dubai's aviation, tourism, commercial and logistics clusters.
- *Dubai Logistics City*, adjacent to the Dubai South-DWC is a free zone with various value-added services and logistics functions, including manufacturing and assembly.

It is worth mentioning that this area continues to expand: not so far from the port, the ongoing megaproject *Dubai Waterfront* was launched in 2007, with the vision 'to create a world-class destination for residents, visitors and businesses'; once completes, it is to become the world's largest man-made artificial island to house 1.5 million people. The Region's first World Expo is also to be hosted by Dubai and the site is located in the Eastern wing of the new airport. The *Downtown Jebel Ali* is also a new mixed-use development—expected to be completed in 2020— that stretches along an 11 km of the Sheikh Zayed Road (between the JAFTZ North and South)—and includes high-rise office buildings and apartment buildings, with access to the main highway, served by a central plaza.

Although, the port-city interface in the Jebel Ali area has been originally development to support the industrial and logistics sector, with recent developments is hosting city functions and therefore attracting permanent residents, as well as the floating business and working population. In fact, the availability of land in the Jebel Ali site had attracted the industries that were formerly located near Port Rashid and the Creek (Dubai's current downtown). While during the mid-80s Port Rashid was the main port of the region, it gradually lost container activities to the new port Jebel Ali and maintained only part of its general cargo flows until the 2008, which was officially closed except for cruise ship operations (Akhavan 2017). Therefore, space was opened in the city centre for property development (residential, office, leisure, etc.), the Dubai Maritime City (a multipurpose maritime zone), and re-developments which resembles the Western post-industrial waterfront project (Hoyle 2000; Marshall 2004).

4.6   Conclusion: Dubai Modal of Port-City Development

To summarize, the four phases of Dubai port-city development can be explained by several factors: (i) the freeport policy, then advanced into several free zones and trade polices (e.g. low taxes and tariffs); (ii) investing oil revenues on capital-intensive trade infrastructures and megaprojects; (iii) an evolution in urban planning approaches, particular land-use (zoning) and urban governance. Despite sharing common features with the Asian consolidation model (Lee et al. 2008; see Chap. 2), Dubai illustrates a particular pattern of port-city evolution. It is interesting to remember that European port-city evolution in Hoyle's (1989) model occurs gradually over several centuries; while Dubai's development, from a small insignificant finishing port—in early 1900—into a global hub-port city by the late 1990s, happened with a relatively faster pace and in less than hundred years.

Figure 4.9 shows a diagram of this four-phase patter of development, with relation to the urbanization factors: the first and second phases mark a relatively mild urban growth, yet the city is investing heavily on infrastructure provision; the results are reflected in the successive phases, when one can perceive the stark rise in the number of population and the urbanized land. Apart from the strategic geopolitical situation

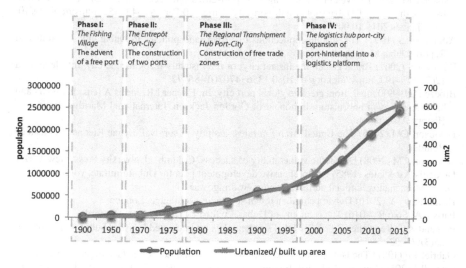

Fig. 4.9 The four phase Dubai port-city development phases with relation to the growth in population and urban and coverage. *Source* Author

of Dubai, this pattern of fast developing port-city, is made possible through a more and more privatized urban governance, as the demand for freehold properties increases, which on the on the other hand leads to a more decentralized urban administration. Conceptualizing the urban development of Dubai as a logistics hub port-city allows us to better understand and analyse the complex characteristics of the rapid growth in the emerging Arabian oil-cities of the Persian Gulf, which are becoming important global players.

Considering its rapid economic success, the explored port-city development model may be appealing for many (fast) developing countries. Earlier, other scholars have also argued the application of 'Dubai model' by other Arab States in this region; imitating strategies for public–private infrastructural investments, real estate expansion through megaproject and industry-specific sectors (Davidson 2008). In 2004, during the international conference for study of traditional environment conference in Sharjah, Elsheshtawy (2010: 250) introduced the term 'Dubaization', or the Dubai model, given the dominance of Dubai at that time and proliferation of its megaproject development strategy in other cities, both regionally in the Middle East—namely Cairo and globally—namely Panama City.

References

Acuto M (2010) High-rise Dubai urban entrepreneurialism and the technology of symbolic power. Cities 27(4):272–284. https://doi.org/10.1016/j.cities.2010.01.003

Acuto M (2014) Dubai in the 'middle.' Int J Urban Reg Res 38(5):1732–1748. https://doi.org/10.1111/1468-2427.12190

Akhavan M (2017) Development dynamics of port-cities interface in the Arab Middle Eastern world—the case of Dubai global hub port-city. Cities 60(Part A):343–352. https://doi.org/10.1016/j.cities.2016.10.009

AlShafieei S (1997) The spatial implications of urban land policies in Dubai city. Unpublished Report, Dubai Municipality

Bagaeen S (2007) Brand Dubai: the instant city; or the instantly recognizable city. Int Plann Stud 12(2):173–197. https://doi.org/10.1080/13563470701486372

Broeze F (1999) Dubai: from creek to global port city. In: Fischer LR, Jarvis A (eds) Harbours and havens: essays in port history in honour of Gordon Jackson. International Maritime Economic History Association

Davidson CM (2005) The United Arab Emirates: a study in survival. Lynne Rienner Publishers, Boulder

Davidson CM (2008) Dubai: the vulnerability of success. Columbia University Press, New York

Doxiadis Associates (1985) Comprehensive development plan for Dubai Emirate, vol 2

Dubai Municipality. Official portal: https://www.dm.gov.ae/

Elsheshtawy Y (2010) Dubai: behind an urban spectacle. Routledge, London

Faris AA, Soto R (2016) The economy of Dubai. Oxford University Press, Oxford

Fernandes C, Rodrigues G (2009) Dubai's potential as an integrated logistics hub. J Appl Bus Res 25(3):77–92

Gabriel EF (1987) The Dubai handbook. Institute for Applied Economic Geography, Ahrensburg

Gornall J (2011) Sun sets on British Empire as UAE raises its flag—The National. Thenational.ae. https://www.thenational.ae/uae/sun-sets-on-british-empire-as-uae-raises-its-flag-1.574721#page5

Government of Dubai (n.d.) Dubai plan 2021: Dubai strategic plan 2015. URL https://www.dubaip lan2021.ae/dsp-2015-2/. Accessed 10 Sept 2017

Harris J (1994) Dubai development plan May 1971. In: Development plans of the GCC states 1962–1995, archive edition

Hawley D (1970) Trucial states. George Allen & Unwin, London

Hoyle BS (1989) The port—city interface: Trends, problems and examples. Geoforum 20(4):429–435. https://doi.org/10.1016/0016-7185(89)90026-2

Hoyle BS (2000) Global and local change on the port-city waterfront. Geogr Rev 90(3):395–417. https://doi.org/10.1111/j.1931-0846.2000.tb00344.x

Jacobs W, Hall PV (2007) What conditions supply chain strategies of ports? The case of Dubai. GeoJournal 68(4):327–342. https://doi.org/10.1007/s10708-007-9092-x

Keshavarzian A (2010) Geopolitics and the genealogy of free trade zones in the Persian Gulf. Geopolitics 15:263–289. https://doi.org/10.1080/14650040903486926

King GR (1997) The history of the UAE: the eve of Islam and Islam period. In: Ghareeb E, Al Abed I (eds) Perspectives on the United Arab Emirates. Trident Press, Abu Dhabi

Krane J (2009) Dubai: the story of the world's fastest city. Atlantic Books, London

Lee S-W, Song D-W, Ducruet C (2008) A tale of Asia's world ports: the spatial evolution in global hub port cities. Geoforum 39(1):372–385. https://doi.org/10.1016/j.geoforum.2007.07.010

Marshall R (2004) Waterfronts in post-industrial cities. Taylor & Francis, New York

Pacione M (2005) City profile: Dubai. Cities 22:255–265. https://doi.org/10.1016/j.cities.2005.02.001

Peck MC (1986) The United Arab Emirates: a venture in unity. Westview Press, Boulder

Ponzini D (2011) Large scale development projects and star architecture in the absence of democratic politics: the case of Abu Dhabi, UAE. Cities 28(3):251–259. https://doi.org/10.1016/j.cities.2011.02.002

Ramos SJ (2010) Dubai amplified: the engineering of a port geography. Ashgate, Farnham

Sigler TJ (2013) Relational cities: Doha, Panama City, and Dubai as 21st century entrepôts. Urban Geogr 34(5):612–633. https://doi.org/10.1080/02723638.2013.778572

Thorpe M, Mitra S (2011) The evolution of the transport and logistics sector in Dubai. Glob Bus Econ Anthol 2(2):342–353

Wilson G (1999) Father of Dubai. Sheikh Rashid bin Saed Al-Maktoum. Media Prima, Dubai

Ziadah R (2018) Transport infrastructure and logistics in the making of Dubai Inc. Int J Urban Reg Res 42(2):182–197. https://doi.org/10.1111/1468-2427.12570

Chapter 5
Economic Diversification, Freight Flows and Transnational Expansion in Dubai Hub Port-City

5.1 Megaproject Development Strategy: Challenges and Policies

Studying Dubai's planning and development phases, one can easily spot the city pattern as a mosaic of megaprojects; modern Dubai has been shaped through dozens of large-scale projects within a very short period of time. With the aim to diversify the oil-based economy, towards hub city making—hub-maritime hub airport, trading-hub, logistics-hub, business-hub, tourists-hub, etc.—the government of Dubai has followed a particular developing approach in constructing cities within the city. Thus, since the 2000s, a property boom was evident, which was then halted due to the aftermath of the global financial crisis in 2008 where some under-construction projects were forced to suspend. Besides, according to the Dubai Strategic Plan 2015, developing large-scale projects has been a driving force in entering the world economy and global city making. By the turn of the twenty-first century and due to the scale of these massive constructions, of some multi-billion-dollar projects, 25% of the world's construction cranes (up to 30,000) were operating in Dubai (in 2006). Functionally, these projects can be divided in three main categories: trade infrastructure (such as seaports, airports), mixed-used urban projects (residential, commercial, etc.) and touristic-based, which are either built from scratch in Dubai's brownfield or reclaimed island developments; mainly financed and administrated by government-led real estate developers (Elsheshtawy 2010), and more recently foreign direct investments.

In the case of Dubai, development through urban megaprojects has been both a cause and response to the global forces. If we consider a megaproject made possible through state interventions, FDI and modernization demands, then these developments, thought local, reflects the global pressure. In the fast-developing countries of the East, in general, and the Persian Gulf States, in particular, the flows of the global and regional capital were invested in massive property development projects. Throughout the formation of a new urban economy, neo liberal political patterns, new urban policy, and the creative cluster, especially in the case of the

© The Author(s), under exclusive license to Springer Nature Singapore Pte Ltd. 2020
M. Akhavan, *Port Geography and Hinterland Development Dynamics*,
PoliMI SpringerBriefs, https://doi.org/10.1007/978-3-030-52578-1_5

fast-developing context, megaprojects can be the drivers of insertion into the global network (Moulaert et al. 2001). However, the rich literature on globalization and global city studies has rather neglected the nature and effect of megaprojects in the formation of a new spatial and economic configuration.

There is no exact and fixed definition of the term 'megaprojects' within the literature, which is mainly defined regarding the specific region and the project in study. In the US cases, some scholars define a megaproject as an expensive and public physical initiative (Altshuler and Luberoff 2003). Flyvbjerg (2005) defines these projects as major infrastructure developments, with a significant cost (more than US$1 billion) and substantial direct and indirect impact on the community, environment and budget. In general, mega projects can be conceived as large-scale, rather complex, capital-intensive infrastructural projects; often linking local, regional and global networks and therefore of great public importance since they, generally, respect the national and regional development strategies. Because of the massive investment and cost overruns, ambitious local development effects and underestimated environmental impacts, there is always a level of uncertainty, especially from a planning point of view. On the other hand, such developments may give rise to socio-spatial fragmentation at the urban level (Swyngedouw et al. 2012). Nevertheless, the government and city leaders mainly deploy the idea of megaprojects for its greater potential economic benefits (Flyvbjerg 2005).

Consequently, the megaproject development strategy for the Persian Gulf region and its fast-developing oil-economy states can be understood as a direct consequence of global pressures, technological development, regional competition in attracting global investment (and FDI), governments dream in global city making, as well as the abundance of revenues brought t by the natural resources (oil and gas). The capital of UAE, Abu Dhabi, has been also seeking apply the strategies to attract footloose investors, tourism and specialized added value functions through mega developments to diversify the oil-economy; megaproject made by star-architects and their spectacular architecture (Ponzini 2011), and therefore emphasising the role of 'archistar' and 'urbanistar' in joining developers and projects (Foster 2002).

In the previous section, while describing each phase, the importance of megaprojects was stressed in several occasions. In 2010, after Shanghai, Dubai was nominated to be the world's largest construction site (Davis 2006). Figure 5.1 illustrates the mega-projects in Dubai, which some have being completed and others still under construction. Here major developments started from the late 1990s and accelerated by mid 2000s. However due to the global economic crisis of the 2008, many of the projects were put on hold and later on some were resumed after 2010. As stated in the latest report of the Dubai 2020 Urban Master Plan, the ultimate carrying capacity of the mega projects (as it was committed before 2008) is estimated to be around 9.5 million inhabitants. This capacity excludes Dubai's existing residential population of 1.91 million inhabitants. These projects are not intended to response to the nationals (local) natural population growth, but to attract foreign investment (Government of Dubai n.d.).

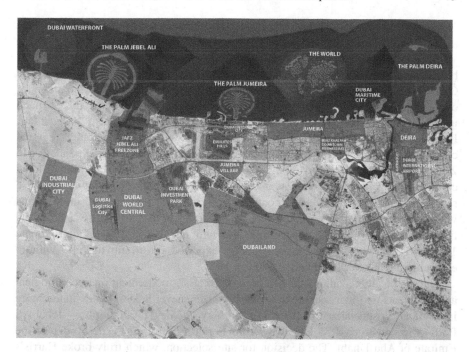

Fig. 5.1 Map of Dubai megaprojects. *Source* Elaborated by the author, base map from Google Earth (year 2015)

5.2 Economic Diversification, Growth in Trade and Urban Development

Many studies on urban development in the Persian Gulf Region (with an exception for Iran and Iraq) have discussed the modern 'Gulf City' and oil urbanization mainly starting from the 1960s (Buckley and Hanieh 2014). In fact, the strategies for economic diversification has been on the agenda of the oil-based Arab states in the Persian Gulf Region—the UAE, Saudi Arabia, Qatar, Bahrain and Oman (El Hag and El Shazly 2012) with the aim to meet the future scarcity of this non-renewable natural source. Nevertheless, the UAE and more specifically Dubai has proved to be most proactive and advanced in this matter (Hvidt 2013), due to some liberalization measures, privatization and opening up towards foreign investments. In the literature, the economic development of Dubai is attributed to some key factors: (i) government-led development and top-down approaches; (ii) fast track decision making; (iii) flexible labour market; (iv) partially neglecting industrialization and focusing on creating a service economy; (v) internationalisation strategies; (vi) promoting investment opportunities; (vii) supply-generated demand (first-mover advantage); (viii) market positioning via branding; (ix) collaboration of international partners and consultants (Hvidt 2009). In line with the central scope of the book, here I shall review Dubai's development path towards making a transshipment hub and logistics centre, focusing on the importance of ports and other trade-ancillary infrastructures in diversifying the economy.

As seen previously, Dubai entered the world of freeports in early 1900s, followed by growing international trade through the expansion of the Creek in late 1950s. The Creek dredging continued in 1960s and 1970s to cope with the rising demand in re-exporting activities; the massive trade-led projects allowed the entrepôt city to expand its urban area southward along the creek's natural urban axis. The construction of Port Rashid in 1972, with 12 berths, was the starting point for Dubai to become a regional container hub port. Due to the growing cargoes, the port's capacity was expanded up to 24 berths in the late 1970s. Although the expansion plans were based on traditional cargo handling, the global forces and improving shipping technologies forced a redesigning strategy to meet the containerization needs. Later on, 13 more berths with a modern container terminal were added, in 1980, to mark Port Rashid as the first and largest modern container port of the region, with a capacity of approximate 100,000 TEUs, with 70% of the containers re-exported—from ship to ship—without even entering the city (Cuthbert 2011).

The era of globalization, 'just in time deliveries', modernization, technological advancements and increasing container ship sizes, nevertheless, necessitated more investments in specialized plants and machinery: the new port complex—Jebel Ali Port—was constructed 35 km away from Port Rashid close to the border of the Emirate of Abu Dhabi. The decision for site selection, which truly broke Harris's ring-radial master plan, was rather political and power related than technical. In fact, with this massive infrastructure project, Dubai leaders intended to mark a strong and competitive border with Abu Dhabi that was decided on the 1968 territorial agreement. Although the project for Jebel Ali new industrial town was never realized, the area (desert) gradually developed into an industrial site, establishing a particular the port-city interface. On the other hand, the new port connected to the historic centre became drivers for a transversal East-West development that forms Dubai's today city pattern as a linear city.

During the first few years of its official opening, the American-based SeaLand Shipping Company was in charge of Jebel Ali Port management and operation; being a competitor to other shipping companies already present in Port Rashid, only a limited number of ships would choose the new port. The location of Jebel Ali Port, far from the city centre where there was nothing but the port infrastructure, was not convenient for the businessmen and the merchants that had their warehouse and offices nearby Port Rashid, which was, on the other hand, becoming increasingly congested by the rising container flows and therefore needed substantial investment for expansion. For these reasons, the new Jebel Ali Port did not perform as expected and, therefore, the Dubai Government was forced to confront the dilemma: two massive ports in proximity and competing for the same business sector as they had two different management systems. To overcome this issue, the government ended the management contracts in both ports and founded a new government-owned, and commercially independent, company in 1991: Port Rashid Authority (PRA) and the Port Authority of Jebel Ali (PAJA) was merged to form the Dubai Port Authority (DPA) (DP World n.d.) The amount of traffic in Port Jebel Ali start to boos and Port

Rashid gradually abandoned its container activities and then officially terminated also its general cargo flows by the end 2000s.

In 1985, the establishment of Jebel Ali Free Trade Zone (JAFZ), with 10,000 ha and adjacent to Jebel Ali Port, was indeed a turning point in Dubai's development into a global transshipment hub. The decision to locate JAFZA, as the pioneering 'free zone' project, alongside the Port was to offer the foreign companies the easy access to unloading facilities, benefited by 100% foreign ownership, and to sustain the new authority at the periphery of the city (Davidson 2008: 115). This zone is also an important factor in encouraging the development towards the south of the city, but also in diversifying the economy through attracting FDI and internationalization. Making reference to the literature on the concept of growth pole and port-regionalization, also argued by Ramos (2010), Jebel Ali Port is a clear example of a growth pole; its site has been planned and developed to infill the growth corridor along the coast, connecting new and expanding port and port-related facilities to the city centre—historical Dubai at the Creek. Figure 5.2 shows a series of views over the port hinterland developing throughput time in period between 1985 and 2015.

Based on data on the amount and type of commodity traffic handled at the two ports of Dubai, type of goods handled (discharged and loaded) are: (i) petroleum oil, (ii) general goods, (iii) container, which in general represents around 70% of the total commodity flow, which confirms the Dubai ports specialization in handling container traffic. Figure 5.3 shows the number of vessels calling at Dubai's ports: in 2015 a total number of 22,524 vessels stopped at Dubai, which grew more than twofold since the mid-1990s.

While analysing the success of a container port, reading the data on the container traffic growth, in terms of the total throughput, can be the basic indicator. Figure 5.4 shows the evolution of container flows, which grew more than 3000 times in 37 years, from 4500 TEUs[1] in 1976 to more than 15 million TEUs in 2015. The compound annual growth rate (CAGR) is also an evident to the sharp increase in its early years of construction (70% in 1980) and then the growth becomes stable in recent years.

5.2.1 Growth in Trade

Extracted from the World Bank Database,[2] the data on world trade shows that trade in Dubai increased by 12 times the global growth, over the period 1994–2009: global exports increased by 120% from US$5 trillion to US$11 trillion, while Dubai enjoyed a 1500% growth in exports valued US$14 billion (Emirates Competitiveness Council 2012). The numbers at the global scale shows Dubai's success in open trade policies and that it is benefiting from globalization and international trade. Figure 5.5 shows the value of import, export and re-export, for the period 1995–2015: the compound

---

[1]Port container traffic measures the flow of containers from land to sea transport modes, and vice versa, in twenty-foot equivalent units (TEUs), a standard-size container.

[2]https://data.worldbank.org/topic/trade.

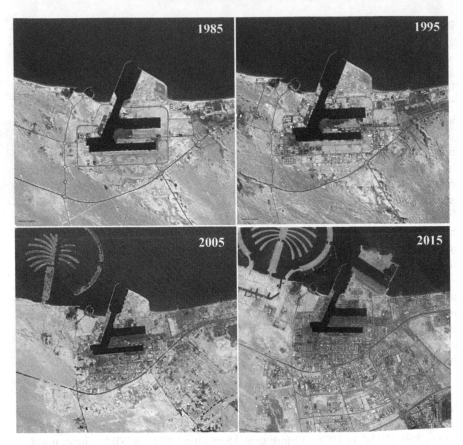

Fig. 5.2 Arial view of the Jebel Ali Port in 1977. *Source* Elaborated by the author, base images are extracted from Google Earth

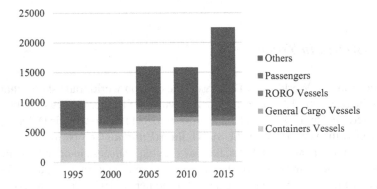

Fig. 5.3 Vessels calling to Dubai's Ports[a] by type. *Source* Own elaboration based on data from Dubai Statistics Center. [a]The data includes both Port Rashid and Jebel Ali Port

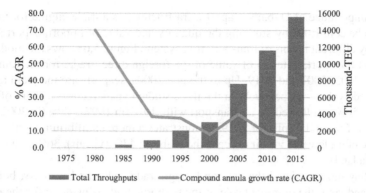

Fig. 5.4 Evolution of container traffic in Dubai Ports[a] (1975–2015). *Source* Own elaboration based on data from Dubai Statistics Center. [a]The data includes both Port Rashid and Jebel Ali Port

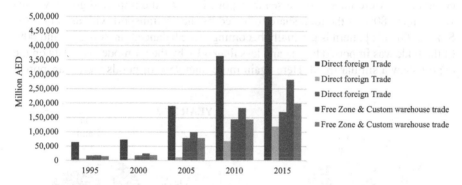

Fig. 5.5 Value of foreign trade by type—value in million AED (1995–2015). *Source* Own elaboration based on data from Statistical Yearbook (2001, 2003, 2005, 2007, 2008, 2009, 2010, 2012)

annual growth rate of the total trade in this 20-year period is calculated as 49%, as the absolute number rose from about 110 billion AED in 1995 to about 1280 billion AED in 2015. Although there was a drop in total trade value during in 2009 global financial crisis, data show that Dubai has recovered in the subsequent years.

Data on the share of import, export and re-export of the total throughput, since its creation till today, show that Jebel Ali Port's relatively high transshipment activities: more than 70% of Dubai's container traffic is re-exported without even entering the city. Reported by the DP World, currently, Dubai is the third largest transshipment centre in the world after Hong Kong and Singapore. Now the question is, whether the increasing level of container transshipment benefits the city? And how? This question can be addressed through investigating the impact of these activities, in terms of number of establishments, generated employment and added value, in the city and specifically in free trade zones.

The importance of Dubai's seaport infrastructure, as a trade engine for the city, can also be confirmed by studying the quantity and value of import, export and re-export by different means of transportation system. Figure 5.6 shows the model split of freight transport, in terms of value of the foreign direct trade, for the four years of 1997, 2002, 2007 and 2019. Up to the late 2000s, seaports are the main anchors of the trade value—respectively in the three studied years records a share of: 78%, 65% and 56%—followed by air transport with, 19% (in 1997), 32% (in 2002), 35% (in 2007). The modal split for 2019 demonstrates a more equilibrium picture, while the share of trade value via air exceeds the seaborn: 47% (via air), 36% (via sea) and 17% (via land).

Reading more in detail the data on foreign trade, value and quantity, bored via sea, air and land, the dominant role of the seaborne trade is evident: for the decade 1997–2007, more than 60% value share of the import, 50% of export and around 50% of re-export is brought by seaports. For the same period, more than 95% of the commodities were imported via sea transport. In 1997, the imported goods via sea valued up to 80% of the total share followed by the air transport with around 20%. Since 2000, this pattern is gradually becoming more balanced: in 2007, around 60% of the trade was imported by sea and less than 40% by the air mode. The data on the export show a similar trend. Here again more than 95% of goods were transported

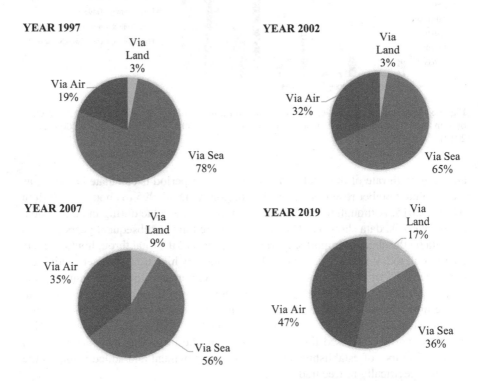

Fig. 5.6 Modal split of transport, based on the value of the total foreign trade for years 1997, 2002, 2007 and 2019. *Source* Own elaboration based on data from Dubai Statistics Center

via sea. However, when it comes to the values exported, the sea covers around 50%, with the other half covered by the other air- and land-based modes, respectively around 30 and 20%. The re-export patterns reveal a slightly different trend compared to the import and export. Although still more that 80% of goods are re-exported through the sea ports, nevertheless the data on the value generated shows a faster trend balance and moving towards a more intermodal model of transportation: in 2007, the modal split of the values generated through re-export were 45% by sea, 35% via air and more than 20% on land. In brief, while the freight movements in and out of Dubai relies heavily on the maritime transport, however according to the compound annual growth rate, with the 20-year period between 1997 and 2019, other modes especially the land and the air transports are showing a growing contribution to the trade: modes of land grew by 415%, air with 188% and the sea transport with 32%. This trend can be explained by the hinterland expansion in the Jebel Ali Area, as described beforehand, more specifically the expansion in the road system and the new Al Maktoum International Airport, part of the DWC project in Jebel Ali area, that launched cargo operation in 2010, has increased the contribution of air transportation to trade.

It is evident that there is a direct relation between the increasing level of freight flows and the value generated by the trade. Taking into account the high proportion of transshipment activities in Dubai, it is important to analyse to what extend these two factors are correlated. Figure 5.7 demonstrates the scatter plot between the two factors of container throughput and the value generated by the foreign direct trade for the period 1997–2019. According to the graph, there is a strong positive correlation when the level of throughput is more than 2.6 million TEUs. Based on the trend line, thus, an increase in, for example, 2.6 million TEUs can generate about 3.4 million AED trade value.

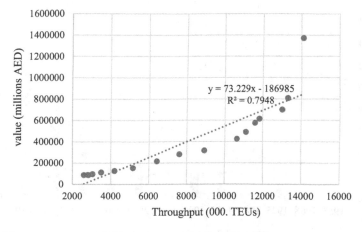

Fig. 5.7 The scatter plot of total throughputs versus value of foreign direct trade in Dubai (1997–2019). *Source* Own elaboration based on data from Dubai Statistics Center

5.2.2 Urban Development Indicators

Studying Dubai's fast urbanization, the first that catches our eyes are the dramatic growth in a relatively short period of time: the numbers raised from 50,000 in 1953 to more than 3.3 million in 2019, which accounts for a 100% CAGR in the this 66-year period. Figure 5.8 expands this data for a 5-year interval, along with the growth rate: though the population shows a constant growth throughout time, the 1970s has enjoyed a relatively higher increase.

Back in the early 70s, Hoyle (1972) discussed the relationship between city size and port throughputs, as two factors depending upon the specific port and urban characteristics within their particular context, to be rather variable and somehow vague. Moreover, Hoyle and Hilling (1970, 1984) argue that along with globalization and advancement in maritime technology in the modern world, the historical strong association between the port and the city has been weakened. In Dubai, the trend is similar to the ones seen in the major Asian cases. As stated beforehand, Dubai is port-city that has been growing rapidly since the 1950s, while the it has maintained and increased its level of throughput until today. Figure 5.9 shows the scatter plot considering total throughputs (TEUs) and population for the period of 1975–2019. The data show a strong correlation between the two indicators: an increase in one million TEUs is associated to a growth of about 220,000 in the absolute number of people. In fact, Dubai is following a similar trend to other global hub port-cities such as Singapore and Hong Kong, in which despite their large population, have kept a relatively high traffic concentration (Lee et al. 2008).

The value and growth of the gross domestic production (hereinafter GDP) is an important factor while analysing an urban development of a city. Although many scholars have discussed the decreasing positive effects of port activities on local

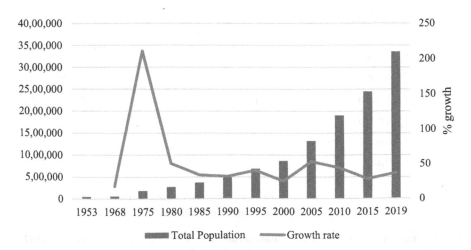

Fig. 5.8 Dubai absolute population and its growth rate (1953–2019). *Source* Own elaboration based on data from Dubai Statistics Center

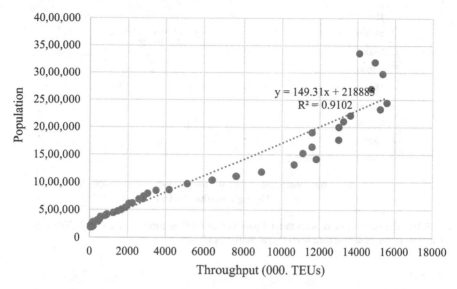

Fig. 5.9 The scatter plot of total throughputs versus population (1975–2019). *Source* Own elaboration based on data from Dubai Statistics Center

economy, ports might generate higher impact on local economies with respect to other infrastructure as they generate externalities on the hinterland (Clark et al. 2004). For the case of Dubai, a simple calculation of data is a prove to the constant growth since 1975–1990 with an annual rise in GDP of about 6%; since 1990 until 2008 the growth is even sharper with more than 10% per year (Scott 2014).

Although the country was hit by the 2008 financial crisis, Dubai was able to get itself back on track with a slightly lower pace with respect to the past: the CAGR in the 10-year period 2008–2018 is calculated as almost 4%. Although academic scholars have long been interested and cautious about the relation between GDP and container traffic, however recent studies show that the ration between global TEU growth and GDP growth is no longer stable (Notteboom 2013). Nevertheless, here, the relationship among maritime container traffic and the GDP (at current prices) is provided in a scatter plot in Fig. 5.10: there is a high positive correlation between traffic throughputs and the level of GDP, for the years between 1997 and 2018, when the throughput is more than 2.8 million TEUs. According to the trend line, the increase in 3 million TEUs is correlated to about 6150 million AED growth in GDP.

As discussed beforehand, the megaproject development has been a government-led strategy towards diversifying the economy, which already led to the point where oil accounts for less than 7% in total GDP (year 2011). Figure 5.11 shows the share of five main sectors to the GDP. From 1997 to 2015 the economy is following a similar trend in which, wholesale and retail trade has been the dominant sector (26.3% of the total GDP in 2015); followed by finance, insurance and real estate (18%). The transportation and manufacturing are the third most important sectors of the economy, with a contribution of around 12%. The overall picture reveals the importance of port

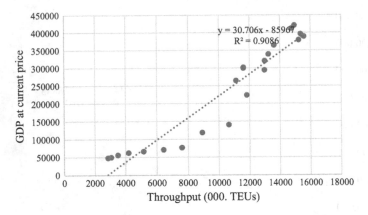

Fig. 5.10 The scatter plot of total throughputs versus GDP at current price (1997–2018). *Source* Own elaboration based on data from Dubai Statistics Center

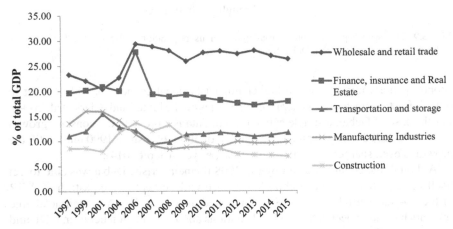

Fig. 5.11 Share of key economic sectors in Dubai's GDP (1997–2015). *Source* Own elaboration based on data from Dubai Statistics Center

cluster (related to trade, transport, manufacturing and partly finance) in contributing to the economic growth (GDP). However deep analysis with detailed-disaggregated data is required to identify the port cluster and calculate accordingly its share in total GDP and then analyse the direct, indirect and induced impact of Dubai's port industry on the local economy.

Some studies have shown that many historical ports, which provided significant number of jobs, after modernizing maritime infrastructures and containerization no longer show a positive employment effect (Grobar 2008; Musso et al. 2000). Nevertheless, other studies discuss that ports with a relatively high transshipment activity might have a lower impact on employment indicators (Bottasso et al. 2013). Considering the lack of disaggregated data on employment, here the focus is on the five main

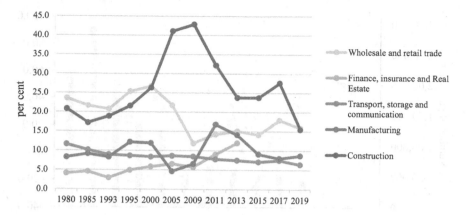

Fig. 5.12 Sectorial distribution of the main economic activities Dubai (1980–2019). *Source* Own elaboration based on data from Dubai Statistics Center

economic sectors. Figure 5.12 shows the sectorial distribution of the main economic activities in Dubai (1980–2011). The booming mega projects since the 1980s led to a dramatic increase in the number of construction workers within the occupational structure: 30% of total economic activities in 2011; however, the construction sector is losing its dominant share: 15.5% in 2019. The 'transport and storage sector' is also showing a declining trend; fell from 11.6% in 1980 to 6.4% in 2019; Manufacturing enjoyed an increase in the share of employment structure in the period 1980–2011, followed by a decreasing number in the recent years: 8.7% in 2019.

Although the focus here is on understanding the role of maritime ports and fright flows in urban and economy development, it is interesting to have a glance at the trend of passenger movement through the two ports of Dubai. As noted previously, Port Rashid, the one based at the city centre, gradually gave-up its container activities to Jebel Ali Port, to mainly function as a cruse port. Figure 5.13 shows the number of passenger flows in Port Rashid and Jebel Ali in the period between 1997 and 2015. The growing number of passengers through the port is also a confirmation to the success in the city's investment on tourism industry as important sector for the government's ambitious economic diversification strategies. In Port Rashid, the number of passengers in this 18-year period grew by 10,000% from about 2800 to 2,900,000 passengers.

5.3 The Spreading System of Free Zones in Developing the Desert: Cities Within a City

The story of Dubai ports as key elements in the development of the city and its trade sector will not be complete without its free trade zone (hereinafter FTZ); in fact, the economy relies heavily on FTZs. The history of FTZ worldwide, in the

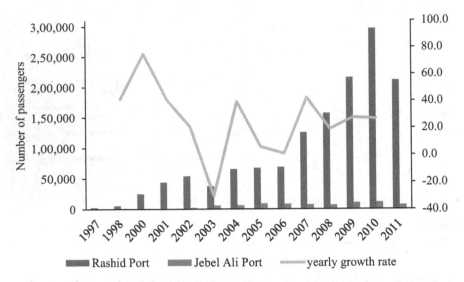

Fig. 5.13 Passenger movement in Dubai Ports[a] (1997–2015). *Source* Own elaboration based on data from Dubai Statistics Center. [a]Here the data considers only Jebel Ali Port and Port Rashid—the other two minor passenger ports of Hamriya Port and Shandagha Port are excluded as the numbers are relatively low and negligible

form of free port, dates back to the Middle Ages. Within the literature there is no unique definition of the FTZ, and given the varied classification of these special zones, it is unlikely to be one agreed by the different disciplines. In general, 'free zones are special areas within the customs territory of the Community where goods are free of import duties, VAT and other import charges' (United Nations 2005: 74). Defined by the Encyclopedia Britannica (n.d.) FTZ is 'an area within which goods may be landed, handled, manufactured or reconfigured, and re-exported without the intervention of the customs authorities'. As seen in Chap. 3 of this book, UNCTAD (1996: 3) has a similar definition for FTZ, developed for the purpose of a port, or a Freeport: 'usually designated area at a port or airport where goods can be imported, stored or processed and re-exported, free of all Custom Duties. It is a free area which normally falls under the authority of the port or airport management'. It is worth underlining that different terms are used to describe the different types of activities performed in zone. For instance, special economic zones (SEZ) or free economic zones, appeared since the late 1950s, are referred to geographically-defined areas, including residential, school, hospital and other facilities, where business, trade and labour regulations are different from the national ones; providing a favourable business environment with several incentives (United Nations 2005). SEZ may include FTZ, free-port, export-processing zone, industrial trade zones, etc. Many studies have discussed the importance of such SEZ as instruments for economic development and liberalization (Chen 1994; Johansson and Nilsson 1997; Romero 1998; Sargent and Mathews 2001). Some preliminary researches have depicted FTZs as a tool for international trade, arguing their advantages for the hosting country (Grubel 1983;

Papadopoulos 1985, 1987). In general, the goal of FTZ is to attract foreign direct investment (hereinafter FDI), and encourage trade and business through favoured tax incentives, fostering industrialization and job creation in the hosting country, which can be managed publicly or privately (see Chap. 3).

The turn of the previous century was followed by changes in the global economy, technological advancement, liberalization of trade, and increase in the number of multinational companies (in production, trade and service). Predominantly, the globalization process rests on rather free and rapid flows of capital, information, labour, people and goods; FTZs are emerging as new spaces of flows within the globalized world. Given the economic basis of the FTZ, developing countries can take advantage of this concept to attract investment and develop a diversified economy, while deploying a neoliberal reform through weakening state's power over the territory and policy making. As an economic development tool, FTZs are mainly adopted in the developing countries (Papadopoulos 1987). Accordingly, this section discusses whether a city with a massive (container) port—a major transshipment hub—should invest on developing and administrating FTZs?

In order to answer this question, it is necessary to trace the FTZs as part of the national and local policies, and to investigate whether and how they contribute to the economic development in terms of number of establishments, generating added value and employment. Here the argument is centred on FTZs in Dubai, particularly the Jebel Ali Free Zone (hereinafter JAFZ): the first attempt to expand the port hinterland, developed as an ancillary infrastructure for the port-related industry. Here, FTZ is considered, on the one hand, as an important planning and policy tool to help boost the economy, and on the other hand as an interface of 'territoriality' and 'transnationalism', in which land is provided with specific regulation on commerce, tax and labour, to allow the rather free flows at different scale (from local to global and vice versa). Although it is not the scope of this research to carry out in-depth survey on Dubai's FTZs, here there is space to introduce the advent of this type of special economic zones and free zones since the 1970s.

Although the fast growing oil-based countries of the Gulf are investing heavily on special economic zones in general and FTZs in particular, a very limited number of studies have considered such zones from a planning and spatial development point of view; and particularly its impact on configuring the transnational flows. Keshavarzian (2010) in his study on the origins of the FTZs in Iran and UAE argues that the creation of such zones at the earlier stage was a response to the creation of a modern city-state through the regional power shift from British imperialism to the American liberal hegemony. Underlining the importance of oil as a determinant factor of power distribution in the city-states of the UAE, the central role of the state in developing the FTZs and controlling the transnational flows has been crucial.

While the UAE has a liberal trade regime, with low tariffs and few non-tariff barriers to trade, yet a number of limitations and conditions are set on foreign investment. Improved market access for its products through multilateral trade liberalization and bilateral and regional trade agreements is a main trade policy objective. Policy formulation and implementation in the UAE takes place both at the federal and the emirate level; the emirates have a relatively high degree of independence. Despite the open trade, the investment regime remains relatively restricted; all investment projects must have 51% domestic capital (WTO-Trade Policy Review 2012).

Among the Arabian states of the Persian Gulf (except Iraq—also known as the Gulf Cooperation Counties—GCC) Dubai was the first to establish trade-oriented free zones as regulatory enclaves with tariff-free imports and full foreign ownership of businesses; often offering far-reaching legal autonomy (Hertog 2019). Since the mid-1980s, Dubai has developed its free zones, which currently accounts for twenty-two zones specialized in varied economic sectors such as trading, logistics, industrial, financial, business services, communication, media and IT, research, healthcare, etc. (UAEstablishment n.d.). Though each zone offers a specialized economic environment, with its relevant authority, what characterizes all the zones are: (i) the provision of world-class infrastructures and facilities by the local government; (ii) tax cuts, if any, and (iii) little regulatory red tape, which are not available in the regulatory domestic area. Moreover, although Dubai's FTZs have administrative bodies, no single authority imposes rules or regulations across all the zones. As illustrated previously in Fig. 5.5, FTZs contribute significantly to the total trade; most recent data (in 2019) show that FTZs hit AED 592 billion, accounting for 43% of Dubai's total trade, generating about 30% of the city's GDP (Dubai Statistics Centre). Moreover, the FTZ development strategy in Dubai has been crucial in attracting, more than 170, global shipping lanes and, up to 120, global airlines (Kumar 2013).

Table 5.1 outlines the chronological list of free zones in Dubai, the existing and under construction, depicting the land cover and the number of establishments (updated in 2015). Accordingly, Fig. 5.14 illustrates the map of Dubai with the location of the main FTZs and industrial areas. Free zone in Dubai is an urban phenomenon; since its advent they have sprawled all over the city in varied locations, from the main city centre towards the waterfront and further inland. From the estimated areas in in Table 5.1, approximately the FTZs spread over 150 km^2, which covers around 20% of the Metropolitan area (the urbanized land covering around 795 km^2) and 4% of the total mainland of Dubai (3978 km^2). The map highlights that the majority of the zones are clustered in the Jebel Ali area, integrated with the (maritime and air) port and industrial areas. This geographical proximity to the main infrastructures provides the foundation to develop the logistics cluster. Based on the fact that majority of the FTZs in Dubai are specialized in trading activities, their strategic location with major transportation infrastructures (sea, air and land) are of great importance.

5.3.1 Jebel Ali Free Zone

Jebel Ali Free Zone—JAFZ is the first and the most important among the free zones in Dubai, which is located at the Jebel Ali area, in between the seaport and the Dubai South Project Dubai (airport and the logistics city). Commenced operations in 1985, JAFZ became the first effort to develop the port hinterland and logistics activities in Dubai's non-urban area, which gave path to the expansion of foreign

Table 5.1 Free trade zones (the existing and under construction) in Dubai

Project	Date of establishment	Land coverage and number of registered companies (2015)	Description of the zone and its specialized activities
Jebel Ali Free Trade Zone (JAFTZ)	1985, expanded in 1990	48,000,000 m^2 6400 companies	The first FTZ in Dubai, JAFZA, located around Jebel Ali Port. Types of activities include manufacturing, processing, assembling, packaging, import/export, distribution, storage, services, etc. Combined with government incentives; a turning point for Dubai's industrialization
Dubai Airport Free Zone	1996	696,000 m^2 1600 companies	Established as part of Dubai government's strategic plan to be an investment driven economy. Today it is one of the fastest growing free zones in the region contributing 4.7% of Dubai's GDP
Dubai Internet City	2000	4,000,000 m^2	Realized for the means of attracting flows of technology
Dubai Cars and Automotive Zone	2000	1,000,000 m^2	DUCAMZ was established for the purpose of re-export of used cars to Asian and African region. Its efficient access to the sea, air and land transportation allows a growing traffic according to the rising demand from the region and beyond
Dubai Media City	2001	300,000 m^2	Established with the aim to upgrade the technological infrastructure

(continued)

Table 5.1 (continued)

Project	Date of establishment	Land coverage and number of registered companies (2015)	Description of the zone and its specialized activities
Dubai Gold and Diamond Park	2001	47,505 m^2 (Phase 1)	The park is an extension of the Jebel Ali Free Zone. Type of company activities includes gold and diamond trading, designing, manufacturing and crafting of gold and diamond jewellery, etc.
Dubai Knowledge Village	Launched in 2003	1,000,000 m^2 Over 400 institutions	An educational trading zone, dedicated to human resource management and learning excellence
Dubai Multi Commodities Centre—DMCC	2002	200 ha 15,000 registered members (year 2018)	Dubai Multi Commodities Centre is a free zone authority for the Jomeirah Lakes Towers (JLT) Free Zone (JTL), with 65,000 living and working. This massive mixed-use zone aims to place Dubai as a global gateway for commodity trade. It is the only international commodity centre in the region, with four main commodity groups: gold, diamonds, pearls and tea

(continued)

Table 5.1 (continued)

Project	Date of establishment	Land coverage and number of registered companies (2015)	Description of the zone and its specialized activities
Dubai Maritime City	First phase completed in 2009—under construction	2,700,000 m^2	A multipurpose maritime zone. When completed, Dubai Maritime City will be a mixed-use development for the maritime industry, comprising industrial, commercial, residential and leisure facilities housed on a man-made peninsula by reclamation of land between Port Rashid and Dubai Dry Docks
Techno Park	Under construction	21,000,000 m^2	Is a fully owned subsidiary of Economic Zones World; which is part of the government owned organisation that developed the JAFTZ and DP World. A knowledge-based technology-centric sustainable business hub, with the aim to make the commercial development of technology
Dubai Silicon Oasis	2004	7,200,000 m^2	An industrial complex located in the heart of Dubai's industrial zone. Fully owned by the Government of Dubai, providing both a living and working integrated community

(continued)

Table 5.1 (continued)

Project	Date of establishment	Land coverage and number of registered companies (2015)	Description of the zone and its specialized activities
Dubai Flower Centre	First phase completed in 2004—under construction	100,000 m^2	Strategically located at Dubai International Airport, the Dubai Flower Centre is a new hub of growth for the floriculture industry. Once fully operational it will serve as an international market with over two billion consumers
Dubai Studio City	2004	2,000,800 m^2 288 companies	As part of Dubai Media city, with the aim to support the growth of broadcast, film, television and music production companies in the Middle East region
International Media Production Zone	2003	4,000,000 m^2	A free zone and freehold area that provides facilities to the media production companies
Dubai HealthCare City	Launched in 2002—Phase II under development	Fase I: 380,900 m^2; Fase II: 1,760,000 m^2	A healthcare free economic zone, to meet the demand for high-quality, patient-centred healthcare
Dubai International Financial Centre—DIFC	2004	445,154 m^2	A federal financial free zone, with its own legal system and courts distinct from those of the wider UAE. The DIFC aims to provide a platform for business and financial institutions of the emerging markets of the region

(continued)

Table 5.1 (continued)

Project	Date of establishment	Land coverage and number of registered companies (2015)	Description of the zone and its specialized activities
Dubai Textile City	Launched in 2007—under construction	460,000 m²	Dedicated to textile industry, the completed zone will function under Jebel Ali Free Zone Authority, to help traders who are mainly in re-export business
Heavy Equipment and Trucks Zone	Launched in 2002—under construction	7,300,000 m²	As part of the efforts to expand the re-export of trucks and heavy equipment from Dubai, HERTZ will be active in re-exports as well as supplying the local market. It operates under the umbrella of the Jebel Ali Free Zone Authority for trading in heavy equipment and trucks
Meydan Free Zone	2009–2010		Located in the heart of Dubai, it is a relatively new business environment that offers infrastructures to enable trading within an international and competitive environment

(continued)

Table 5.1 (continued)

Project	Date of establishment	Land coverage and number of registered companies (2015)	Description of the zone and its specialized activities
Dubai South (formally known as Dubai World Central (DWC))	(Under construction) The Airport was inaugurated in 2010	140,000,000 m^2	Planned as a purpose-built project (airport city), Dubai South (ex DWC) initiative aims to fulfil the emirate's vision to become a leading international trade centre. It includes several key components: the Al Maktoum International Airport (AMIA), Logistics District, Aviation District, Business Park, Commercial District, Residential District, Golf District, Exhibition District and Humanitarian District—which collectively serve as a strategic platform for the expansion and growth of aviation, logistics, light industry and ancillary service businesses. The 'City of you' is to house a million population

Source Own elaboration, gathered from several online sources, such as https://www.uaefreezones.com/

trade. As stated by the chairman of the DP World: '*Jebel Ali Port plays a pivotal role in enabling international trade so companies operating in JAFZA can import and re-export their goods and products to the various countries of the region*' (Xinhua 2017a). Furthermore, some of the other free zones are an expansion of the JAFZ such as Dubai Technology Park, Silicon City, and the industrial area. Although other FTZs, such as Dubai Internet City and Media City, are of great importance due to its success in attracting technological flows and contributing to development in the knowledge-based economy, however JAFZ continues to be the dominating attraction of foreign trade among other zones (Dubai Chamber of Commerce 2004).

The mega project for Jebel Ali Port, constructed in 1979, was planned as part of a greater trading and industrial complex (including a dry dock and aluminium

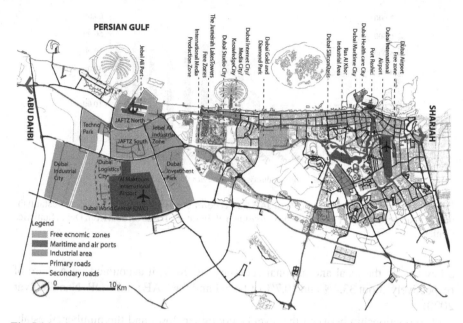

Fig. 5.14 Map of Dubai's free trade zones (the existing and under construction)—year 2015. *Source* Own elaboration based on several sources

smelter) which then in 1985 was expanded into an approximate 50 km² free trade zone. Taking advantage of the strategic location and modern infrastructures, the Jebel Ali port and the adjacent free zone has been preliminarily functioning as the transshipment hub of the region. Hence there is a strong reciprocal relation between the Port and the free trade zone. The main function of JAFZ includes, port discharge cargo, warehousing and logistics, trade, manufacturing and service for the local, regional and international players; commodities are mainly re-exported after being assembled, labelled and packaged; in fact, 74% of trade is re-exported (Ziadah 2018). JAFZ started with 19 companies in 1985 and showed a rapid growth soon after the opening with about 300 diverse companies. This number was raised to 2000 establishments at the early years of the 2000s, and to 6500 companies 2010, most recent data record 7500 companies and 135,000 workers (Nandkeolyar 2020). The multi-scalar market at JAFZ facilitates from the massive container port of Jebel Ali, which serves more than 150 shipping lines connecting Dubai to about 115 direct ports of call worldwide; it is a crucial enabler of growth in the UAE.

Figure 5.15 shows the number of establishments at JAFZ based on the three main economic activities for the period 1997–2011. Since 1997, more than 75% of the business are dedicated to trading, warehousing and distribution, while less than 20% in industry and the rest in services, which represents a rather steady growth. As discussed previously, free trade zones are preliminary established to attract FDI and boost the international trade. In this regard, JAFZ has been responsible for almost 24% of total FDI in 2018 and generated trade worth of $93 billion. It also plays

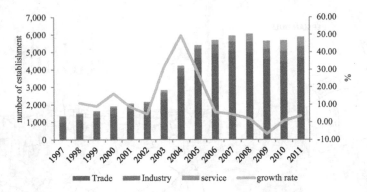

Fig. 5.15 Number of establishments in JAFZ by type of economic activity versus the establishment's yearly growth rate (1997–2011). *Source* Own elaboration, based on data from Dubai Statistics Center

a key role in the local and national economy: in 2017, it accounted for a share of respectively about 33.4% and 10.7% of Dubai and the UAE total GDP (Nandkeolyar 2020).

The relationship between the level of container flows and the number of establishments (for the years between 1997 and 2012) is illustrated in a scatter plot in Fig. 5.16: the graph shows a high correlation between the two factors. According to the trend line, the increase in 1 million TEUs is correlated to the 500 increases in the number of establishments.

Figure 5.17 shows the number of establishments in JAFZ based on the place of registration, into four groups of: Emirates (locals), Gulf States (GCC), Arabic countries, and others. Since 1997, more than 80% of the companies were non-Emirates. However, there is a growing balance between the presence of the Arab (including the locals and non-locals) and non-Arab traders in the Jebel Ali Free Trade Zone:

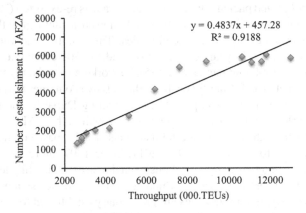

Fig. 5.16 The scatter plot of total throughputs versus number of establishments at JAFZA (1997–2012). *Source* Own elaboration, based on data from Dubai Statistics Center

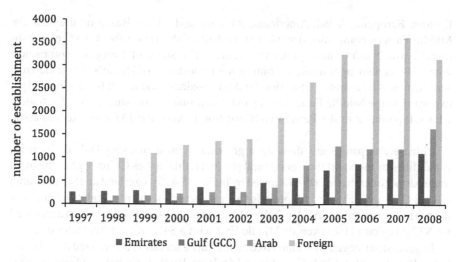

Fig. 5.17 Number of establishments in the Jebel Ali Free Trade Zone by the place of registration. *Source* Own elaboration, based on data from Dubai Statistics Center

from the total number of 6035 establishments in JAFZA (in 2008), the local emirates registered 18%, the GCC countries of 2.5%, other Arab world 27% and the rest with 52% are the multinationals (non-Arabic) companies. This numbers are a confirmation to the success in attracting foreign companies due to the benefits offered by the hosting city-state. The multinationals can use Dubai free zones for manufacturing, warehousing and/or distributing without having to burden the import duties or partnership with the local companies.

Figure 5.18 illustrates the share of multinational companies established at the JAFZ in 2006 based on the region of their origin in five major groups of Middle

Fig. 5.18 Geographical distribution of establishments at JAFTZ. *Source* Jebel Ali Free Zone Association corporate brochure 2007

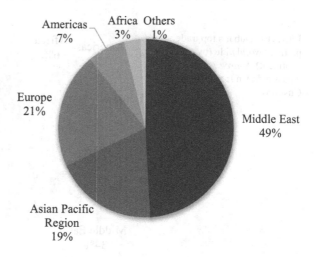

Eastern, European, Asian, Americans, Africans and others. Based on the data, the Middle Eastern companies occupied about half of the zone (about 3000 establishments), which shows its regional importance. The share of European companies, with 21%, is also significant, to confirm the importance of Dubai's free trade zone for Europeans as a strategic location for doing business and trade; for storing and re-exporting to the Middle East, African and Asian market. In other words, for traders who sell products in the Persian Gulf and South Asia, the JAFZ is a dream come true.

Chinese companies are showing a growing interest in using Dubai's logistics capabilities to re-export their goods and products; this makes China, by far, the main trade partner of Dubai in 2019, followed by India, USA, Switzerland and Saudi Arabia (Dubai Customs). Figure 5.19 provides a clear picture of the regional distribution of the top trade partners of JAFZ in 2016. The trade with Asian Pacific, which accounted for $32.4 billion (41%) exceeds Middle East with a 34% of the total value share.

In general, strategically located free zones can be served as centralized distribution hubs (Papadopoulos 1987; Tansuhaj and Jackson 1989; Tansuhaj and Gentry 1987; Mathur 1993). This is also valid for the case of JAFZ, with a rising number of international firms using the zone as a significant distribution hub for commodity movement to the neighbouring market. Outside the JAFZ, the federal commercial law requires foreigners to go into business with national UAE firms or sponsors who are entitled to own at least 51% of the business and its profits. But inside the zone, foreign companies are free to manufacture, trade and form joint ventures.

Most companies are set up at JAFZ for the purpose of re-exporting goods through Dubai to other markets, the largest being Iran and India, given the following as it provides advantages policies for the investors: (i) 100% fully owned foreign investment, is not subject to the provisions of the company law of foreign investment which accounts for 49%, domestic 51%; (ii) the foreign company enjoys exemption from income tax for 50 years, which may be extended for 15 years after the expiration of the period; (iii) 100% repatriation of capital and profits; (iv) the goods of

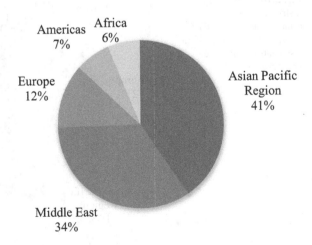

Fig. 5.19 Dubai's top trade partners worldwide (by continent). *Source* Author, data are drawn from Dubai Customs

import and export exempt entirely and does not levy income tax; (v) the registration procedure is simple, without cumbersome bureaucratic procedures and no minimum capital investment restrictions; (vi) the recruitment procedure is simple, efficient and no limit for employers; (vii) low-cost energy supply; (viii) excellent infrastructure, support services and communications; (ix) access to a customer market of 1.4 billion people (Jebel Ali Free Zone Authority n.d.).

Moreover, there is a set of land management policies to protect the national security of land yet develop the land value: the leased land cannot be sold; the enterprise has the authority to use the land for building workshop, office or warehouse. In any case, the rent of the land, office, warehouse and other hardware facilities is a major source of income for zone management mechanism. In order to attract foreign investment, JAFZ has a series of special employment policies: simplifying the visa application procedures; reducing the visa fee, provide certain free work visas according to the investment and the scale of enterprises. With the vision to be the leading provider of sustainable industrial and logistics infrastructure solutions, JAFZ until today has played a pivotal role in Dubai's trade, industrial and logistics development. According to the data provided by the JAFZ authorities, the zone has generated 135,000 jobs while attracting more than 20% of the UAE's foreign direct investment (Jebel Ali Free Zone Portal). As affirmed by DP World Chairman *'the value and volume of trade through JAFZA underlines the strength of the national economy and its ability to adapt to global trading conditions, create investment opportunities and open up new markets to exports from the United Arab Emirates'* (Xinhua 2017a).

Dubai rulers have taken advantage of 'zoning' as the preliminarily strategy to respond to the global forces, geopolitical shifts and the emergence of competitors. On the other hand, the idea of creating zones towards trading-hub making is correlated with the Persian Gulf's long history in trade, which has long connected the Arabian Peninsula and Iranian Plantae to the Indian Ocean. The particular case of JAFZA, with its industrial basis and proximity to the port(s), designed to become a new growth pole, away from the historical city centre, has created a new pattern of port-city interface. Considering the JAFZ as the interface between the port and city, it is important to underline how the port and the city.

From the study on JAFZ, through establishing a free zone adjacent to a hub transshipment container port, the following features can be achieved: (i) to generate economic activities and employment, (ii) to develop the trading hubs and to increase the regional trade, (iii) to develop the local industry, by encouraging the logistics and manufacturing firms specialized on export and re/export activities, (iv) to attract foreign investors as well as foreign highly skilled employees and technologies, (v) to develop the infrastructural basis of the city-region.

5.4   Making an Integrated Logistics Platform at Dubai's Port-City Interface

The logistics cluster in Dubai has always been a key driving force for economic activities, which accounts for about 12% of GDP (in 2015) and 8% of employment (in 2019). The logistics industry—compose of transportation, inventory, material handling, packaging, warehousing, wholesale and retail—is a determining factor in the competitiveness of the economy. As a response to global demand and following the hub making strategy, the project of *Dubai Logistics Corridor* was officially inaugurated in 2010; with the concept of linking sea, land and air, it spreads over an area of approximately 200 km^2 (WAM Dubai 2010). Figure 5.20 illustrates the

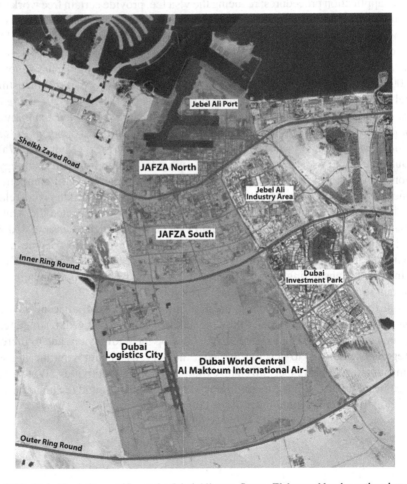

Fig. 5.20 Dubai logistics corridor at the Jebel Ali area. *Source* Elaborated by the author, baseman from Google Earth (year 2015)

area, which encompasses: Jebel Ali Port and Free Zone; airport city within the Dubai South (former DWC); industrial area and the logistics city. The Jebel Ali logistics based platform is part of the government's strategy in planning 'cities within the city' mega projects (Bagaeen 2007). As noted beforehand, this planning approach has been replicated in other part of the city with more specialized free zones, to attract and accommodate new sectors with specific infrastructures (Davidson 2008: 116), namely Dubai Maritime City, Dubai Internet City, Dubai Media City, etc. The connection between varied modes of transport, more importantly maritime and aviation hub ports, coupled with soft infrastructure and a favourable business and investment environment ensured in free zones, make Dubai—for specifically JAFZ— a highly attractive base for global and local logistics firms (Ziadah 2018).

Since the discovery of the oil in Dubai, the expansion of trade-oriented infrastructure continued. Planning the integrated logistics platform, with the vision to host around 10,000 companies with 305,000 employees, can be considered as an important step in shifting from a transshipment hub towards a more specialized logistics hub-port within the region. This synergy between the local, regional and global economy would boost the competitiveness at the regional level, while redefining the port and city relation at the local level. As seen in the previous chapter, the Jebel Ali area was preliminary planned to become an industrial town, with Dubai to remain as the administrative and urban centre. Although the Jebel Ali project was never realized as a new town, the Port and the adjacent Free Zone formed the industrial base. The megaproject Dubai South, formerly known as DWC, as an airport city, merges with the existing port, industrial zones and JAFTZ to create a kind of the integrated logistics platform at the port-city interface.

Due to the rising freight flows and demand for trade, apart from the provision of trade-oriented infrastructures, which started since the 1960s, Dubai government has enhanced its institutional framework. In fact, state plays a key and active role in the development of Dubai's logistics cluster. As stated earlier, the authorities of the Port Rashid and Jebel Ali Port merged in 1991 to form the Dubai Port Authority— DPA. In 1999, Dubai Ports International—DPI was recognized as an international port management company, which later in 2005 joined DPA to form the Dubai Ports World (DP World) (see Sect. 5.6).

An important step towards enhancing the integration of port to the logistics and global supply chain, in 2001, was to merge DPI with JAFZA, Dubai Customs Department, and DP World to create the Ports Customs and Free Zone Corporation (PCFC) (see Jacobs and Hall 2007). Prior to 2000s, trading had a complex and fragmented bureaucratic process,[3] which was a threat to the ease of doing business and the potential growth in the global supply chain. Dubai based entity of PCFC's effort

---

[3]From a trader's perspective the process was bureaucratic, cumbersome and time-consuming. To ship goods businesses had to interact independently with numerous government entities such as customs, health authorities, licensing entities, free zone and sea or airport services to name a few. They were required to submit an array of paper documents in person, often for overlapping administrative requirements. Businesses also had to engage the services of a host of private firms including freight companies, trucking companies, shipping lines and brokers and the like (Emirates Competitiveness Council 2012).

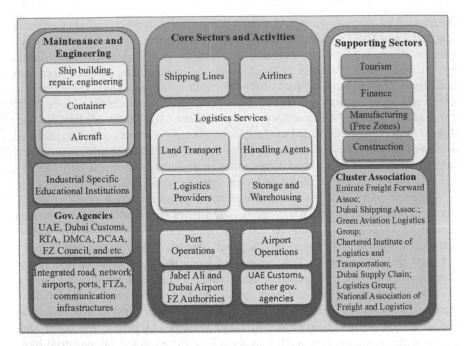

Fig. 5.21 Dubai cluster map of multimodal trade, transportation and logistics sector. *Source* Adapted from Afrikian (2017)

was to create an independent network, through merging the relevant government and private entities in a merged trade process. Figure 5.21 shows Dubai's cluster map and the complex set of public and private institutional actors that creates Dubai multimodal trade, transportation and logistics cluster. With the aim to consolidate Dubai global competitiveness as a multimodal transportation and logistics hub, in 2011 the Department of Economic Development, Government of Dubai, launched the cluster platform (Albawaba 2011).

5.5 Inserting into the National Context and Consolidating a Hub Position in the Region

5.5.1 Situating Dubai Within the Nation-State

Since its creation, United Arab Emirates—UAE, has enjoyed a rapid urbanization trend and population growth from approximately 550,000 in 1975 to more than 9.5 million in 2018; roughly 35% of this population reside in Dubai, which makes this federal state the most populated among the seven emirates. The nation also marks a remarkable economic development, taking into account the GDP growth rate of more than 450% over a 38-years period (1980–2018), from about US$75 billion in 1980 to US$414 billion in 2018 (source of data: International Monetary Fund—IMF and World Bank). This makes UAE the second largest economy among the GCC countries, after Saudi Arabia. Figure 5.22 shows the GDP growth in UAE and Dubai for the period 1997–2018. Dubai with a more diversified economy which is less dependent to the oil has enjoyed a higher growth rate. Since the 2000s Dubai accounts for more than a quarter of the national GDP.

In a globalized economy and through the dramatic rise in the international commodity flows, the trade and logistics sector become one of the key economic activities. The world exports have increased by 120% from US$5 trillion to US$11 trillion, in 15 years between 1994 and 2009, exceeding the world GDP growth (World Development Indicators). According to the World Back Business Report, UAE is ranked 5th among 183 countries, for ease of Trading Across Borders (World Bank 2012). This is a proof to the country's success in emerging as a preferred trade and logistics hub. This status is highly dependent on Dubai.

Based on the raw data provided by the government, natural resources still comprise a great share of the national GDP, though the economy is moving towards diversification: in 2018, the sector 'mining and quarrying (includes crude oil and natural gas)' contributes to more than quarter of the total GDP. In this regard, Fig. 5.23 shows a broader picture of the sectorial distribution of the main economic activities to the total GDP for the years between 1975 and 2017. UAE has recently experienced improvements in its non-oil sectors. After the oil and gas sector, for the year 2017, 'finance, insurance and real estate' (16%), 'wholesale retail trade and repairing services' (12.3%), 'manufacturing industries' (8.8%), 'construction' (8.7%) and 'transport, storage and communication' (5.9%) are the main economic activities. The two sectors of 'manufacturing industries' and 'finance, insurance and real estate' have had the highest growth rate within this 42-year period. Although natural resources yet mark a high share of the national GDP, yet the growth rate in

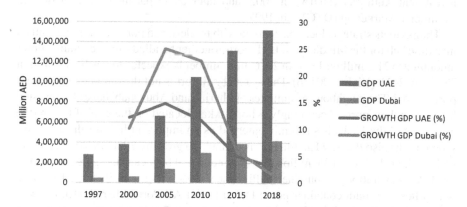

Fig. 5.22 UAE and Dubai GDP (at current price-million AED) (1997–2018). *Source* Own elaboration, based on data is gathered from Dubai Statistics Centre and Federal Competitiveness and Statistics Authority

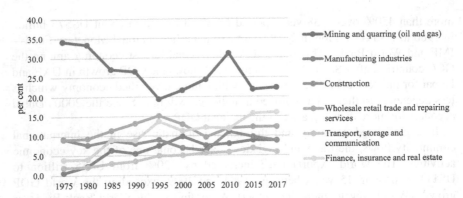

Fig. 5.23 National sectorial distribution of the main economic activities (1975–2017). *Source* Elaborated by the author—data is gathered from the official data portal of the UAE Government https://bayanat.ae/

other sectors depicts that the economy, overall, is moving towards diversification. Dubai is the only emirate with a diversified economic structure, which currently the oil contributes with less that 2% to its GDP. Yet in other emirates, the economy is still highly dependent on the oil; for instance, in Abu Dhabi the oil and gas comprise around about half of the its total GDP.

From an institutional point of view, though Abu Dhabi has been the Capital and hence hosting all ministries to govern the major domestic and external national affairs, yet a number of governmental institutions were established in Dubai, which have been playing a significant role in the Emirate's development. The most important are the Dubai Municipality, already recognized in 1957, followed by the Dubai Commerce and Tourism Promotion Board in 1989, Dubai Ports Authority (DPA) in 1991, the Department of Economic Development (DED) in 1992, the Dubai Development and Investment Authority (DDIA) in 2002, and later on Department of Tourism and Commerce Marketing (DTCM) in 1997.

Thanks to its strategic location coupled with modernized ports, UAE is currently a maritime hub for the Middle East. UAE ports together handled a total container flows amounted to 21.3 million TEUs in 2016; this number is expected to reach 28.4 million TEUs by 2021 (Xinhua 2017b). Dubai Jebel Ali Port is not the only large container port in UAE. The neighbouring emirates of Sharjah and Abu Dhabi have also invested heavily on their ports. Accordingly, Khor Fakkan located on the gulf of Oman (the east coast of the UAE) has recently emerged as an important hub for the east-west global trade; also the new Khalifa Port is planned to grow by 2030 up to the capacity of 15 million TEUs and 35 million tons of general and bulk cargo. Figure 5.24 shows the UAE's evolution of container traffic since 1979 along with its three Emirates, which host the main container ports: Dubai (Jebel Ali Port), Sharjah (Khorfakkan Port) and Abu Dhabi (Mina Zayed Port). In the late 1970s Dubai handled more than 95% of the national traffic, yet the percentage change shows that other two ports are

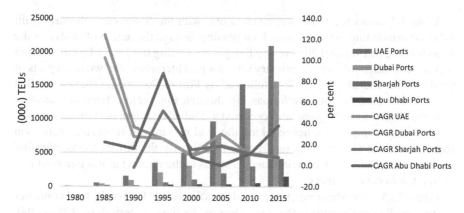

Fig. 5.24 Evolution of the container traffic in UAE and the three main Emirates of Dubai, Abu Dhabi and Sharjah (1980–2015). *Source* Own elaboration based on varied open-data sources of the official statistics department

also developing fast to cover together around 25% share of the containers entering UAE.

5.5.2   The Maritime Network Structure of the Middle Eastern Major Ports

The story of the changing position of Dubai ports within the regional maritime network, and eventually becoming the determining node—a hub-port—is quite interesting. Here the analysis is focused on a part of the Middle Eastern region, the Arab world, which includes the four sub-regions of Persian Gulf, Arabian Sea, Gulf of Oman and Red Sea; only the main container ports (above 100,000 TEUs in 2011) of each country is considered. However due to the lack of data, some ports were removed from the analysis (e.g. Iraq's ports).

Container ports worldwide are competing to expand their capacity to keep pace with the needs of the fast-growing trade requirements (regional shipping and port development). Ports of the Middle Eastern region, especially those favoured by the oil revenues, are also pursuing a port development strategy. As reported by the website thenational.ae, the GCC countries are investing their oil revenues on further developing ports along with free trade zones and industrial complexes with the aim to promote trade, create employment and diversify the non-oil-based economy (Saadi 2013a). As the regional trading hub, the UAE ports, with Port Jebel Ali taking the ·lead, has invested $8.6 billion, by 2013, on expanding their port-related facilities. Thanks to its port and anchored infrastructures, in 2012, UAE accounted for 50% of the GCC imports of $444 billion and 33% of the exports worth $1.06 trillion (Saadi 2013b).

Table 5.2 shows the main port-cities, along with the Share of each port (Unit: %TEUs) in handling the total throughput passing through the region of study, for the periods between 1980 and 2011 (every five years). During the 1980s, by considering only the container volume as a primary filter for port hierarchy, the two main ports of Saudi Arabia, Jeddah Port (43.7%) followed by Port of Dammam (19.5%) handled the most share of the total traffic passing through this region. However, starting from the 1990s onward a great shift occurred as Dubai's modern port infrastructures become more competitive, hence taking the lead by far with recording more than 40% (13 million TEUs) of the total traffic in 2011. By the turn of the twenty-first century, not only Dubai has maintained its status in the region but has emerged into the top ten world containers.

Figure 5.25 shows the compound annual growth rate (CAGR) of the total container volume handled by the ports in the studied region, for the four periods of: 1980–1990, 1990–1996, 1996–2006 and 2006–2011. Four main categories of fast, moderate, slow and negative growth (the darker colour shows a higher growth rate) can be identified. Since the late 1980s, the general trend is that UAE ports are growing rapidly and overtaking their former competitors (i.e. Jeddah, Dammam and Kuwait). It appears that, from the mid-1990s, Dubai's regional rivals, more importantly Jeddah, Sharjah, Salalah and Dammam, are growing fast to boost the regional challenge and potentials.

In order to analyse a port's changing position and degree of hub-ness within the maritime network, the traffic flows in terms of the total throughput, as the principle indicator in measuring the port performance, is indeed not enough. Hence, a network analysis is essential to explore Dubai's development as a transshipment hub within the inter-port network of the region, based on the total amount of throughputs passing through each port-city, and also number of calls to the specific port. Within the extensive research related to port performance and inter-port relations, there is still a gap in considering a network-oriented approach while positioning a port in the relative network. With reference to the conventional methods of network analysis in transport geography, Ducruet et al. (2010) and Ducruet (2009) have made an attempt in measuring the changing network position of ports by deploying data on inter-port flows, with a focus on mainly the Eastern Asian and Western regions. Although Middle East is a strategic region and its ports are becoming more connected into the world network, yet scholars have largely neglected this region and we know little about the port structure of this part of the world.

In this regard, the research study by Akhavan (2015) has analysed the inter-port linkage for the Middle Eastern ports based on data on vessel movements and capacity (containers) for the years 1996, 2006, and 2001 (month of May). Taking into account the dynamic of competition among ports, different trajectory of network system can be recognized. This study further discusses that the general structure of the maritime network in each year is following a similar trend, with Dubai strongly maintaining and enhancing its' position, not only in attracting more throughputs but also in terms of the number and volume of direct calls. While the neighbouring ports are growing rapidly to mark a place into the system, it seems that Dubai's benefiting from their position—as feeders—in becoming a stronger transshipment hub of the region.

Table 5.2 Share of main container port-city in handling the total throughput passing through the Middle East (unit percent TEUs, 1980–2011)

Regions	Port	Port-city/region	Country	1980	1985	1990	1995	2000	2005	2011
Persian Gulf	Dubai	Emirate of Dubai	UAE	4.9	18.4	34.0	39.2	33.2	38.5	42.4
Red sea	Jeddah	Makkah Province	Saudi Arabia	43.7	32.0	20.4	15.0	11.3	14.3	13.1
Arabian sea	Salalah	Dhofar Governorate	Oman	NA	0.09	0.02	0.01	11.2	12.6	10.4
Persian Gulf	Khor Fakkan/Sharjah	Emirate of Sharjah	UAE	2.37	1.7	6.9	12.0	12.1	10.9	10.5
Persian Gulf	Shahid Rajaee/Bandar Abbas	Hormozgan Province	Iran	NA	NA	NA	3.2	4.5	6.5	9
Persian Gulf	Dammam (King Abdul Aziz)	Eastern Province	Saudi Arabia	19.5	12.0	8.6	5.3	4.93	4.5	4.9
Persian Gulf	Shuwaikh	Al Asimah Governorate	Kuwait	13.2	11.4	4.6	4.1	3.37	3.4	NA
Red sea	Aqaba	Aqaba Governorate	Jordan	3.3	5.1	3.1	2.1	2.62	2	2.3
Red sea	Aden	Aden Province	Yemen	0.1	0.2	0.3	0.2	2.69	1.6	0.5
Persian Gulf	Abu Dhabi	Emirate of Abu Dhabi	UAE	0.9	1.2	1.7	4.6	3.69	1.7	2.5
Red sea	Port Sudan	Red Sea State	Sudan	5.8	0.8	1.3	NA	1.02	1.6	1.5
Persian Gulf	Mina Salman	Manama	Bahrain	4.6	4.9	2.8	1.9	1.92	1.3	1.2
Gulf of Oman	Fujairah	Emirate of Fujairah	UAE	NA	6.3	15.4	10.5	5.87	0.3	NA

Source Adapted from Akhavan (2019: 187)

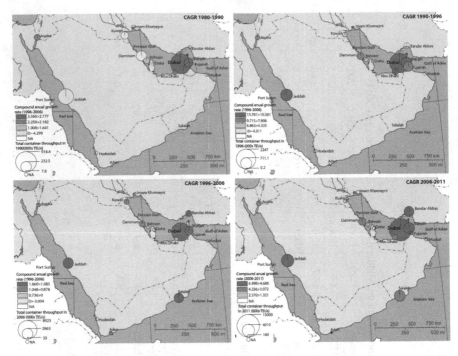

Fig. 5.25 CAGR of total container volume handled at ports in the region of study (1980–1996), (1990–1996), (1996–2006), (2006–2011). *Source* Own elaboration based on the varied opendata sources—governmental statistics centres

5.6 International Expansion and Configuring the Transnational Flows: Dubaization of Port-Hinterland?

A preliminary factor for a success of a port, in attracting traffic flows, may be attributed to its location with respect to the global maritime freight transport system, which includes north-south a and east-west routes. Dubai is strategically located at the crossroad of the major world trading routs between East and West, and world-wide maritime system. Dubai's port development initiatives have shown a constant increasing level of traffic flows, and hence emerged among the top ten major global ports since the early 2000s. Table 5.3 outlines the hierarchy of the top ten ports world-wide based on the total containers handled. Chinese ports are dominating the global container throughput: seven of the top ten world busiest container ports are Chinese base; Singapore has always kept its second position. Being the only Middle Eastern example in the ranking, with an annual growth rate of more than 20%, Dubai port position has improved from 13th in 2000 to the 10th in 2018. Although the European region historically has hosted important ports, only Rotterdam (14.5 million TEUs),

Table 5.3 The top ten world port hierarchy. Year 2018 versus 2000

Rank 2018 (2000)	Port	Country	Volume 2018 (2000) (million TEUs)	CAGR (2000–2018) (%)
1 (6)	Shanghai	China	42.01 (5.61)	36
2 (2)	Singapore	Singapore	36.6 (17.04)	6.4
3 (11)	Shenzhen	China	27.74 (3.99)	33
4 (68)	Ningbo-Zhoushan	China	26.35 (0.90)	175
5 (38)	Guangzhou Harbor	China	21.87 (1.43)	79.4
6 (3)	Busan	South Korea	21.66 (7.54)	10.4
7 (1)	Hong Kong	China	19.60 (18.10)	0.5
8 (24)	Qingdao	China	18.26 (2.12)	42.3
9 (32)	Tianjin	China	16.00 (1.71)	46.4
10 (13)	**Jebel Ali Port**	**UAE**	**14.95 (3.06)**	21.6

Source Elaborated by the author, data from the Portal of World Shipping Council: https://www.worldshipping.org/
The bold values highlight the 'Jebel Ali Port' (the main port of Dubai and Emirates, and the focus of this chapter) being the only Middle Eastern port among the top ten work port hierarchy: 21.6% increase in its compound annual growth rate from the year 2000 to 2018

Antwerp (11.1 million TEUs) and Hamburg (8.7 million TEUs) are seen among major global ports, respectively ranking 11th, 13th and 19th in 2018.

Although it is not the scope of this book to go in deep into studying Dubai's port operator internationalizing strategies, the key actions are nevertheless highlighted to understand the importance and power of this port-city as a global player in ports overseas while controlling the commodity flows worldwide. The formation of a wholly owned subsidiary Dubai Ports International—DPI in 1999 along with its first foreign project Jeddah Islamic Port in Saudi Arabia, in collaboration with a local market, in same year, are the first attempts made to expand its presence beyond local borders. One year later, DPI took control over Port of Djibouti and Djibouti Airport, as well as managing its marine, bulk and container operations, and logistics zone. The global footprints expanded by taking over Visakhapatnam Port (India) in 2002, Constanta (Romania) in 2003 and Cochin (India) in 2004. An important step in global developing strategies was taken in 2005 when DPI buys the North-Carola based CSX World Terminal LLC, which was a leading global container terminal operator with key strategic assets in some of the world's fastest growing markets, including Asia and South America. This acquisition makes Dubai among top six world operators of terminals; several ports were hence added to its portfolio from the Middle East, Turkey and China. Later in the same year, another important step is the creation of DP World,[4] through emerging DPA and DPI as an Emirati marine terminal operator—wholly owned by the government—which is based in Dubai and Jebel Ali is its

---

[4]DP World is an Emirati maritime based terminal operator, which is based in Dubai and Jebel Ali is the its flagship project.

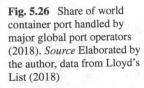

Fig. 5.26 Share of world container port handled by major global port operators (2018). *Source* Elaborated by the author, data from Lloyd's List (2018)

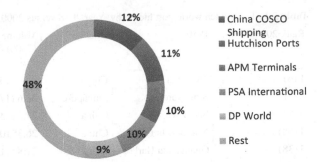

flagship project. The rapid global expansion of DP World started in the same year of its foundation by the acquisition of the world's fourth-largest port operator: The Peninsular and Oriental Steam Navigation Company (P&O) for $7 billion. In order to maintain its antagonistic position in the top global leaders of supply chains, DP World follows a competitive strategy including innovation, capacity, sustainability and funding. According to the company's data, more than 70 million TEUs were handled in 2018, placing the company among the top five worldwide in maritime terminals by throughput (DP World Annual Report n.d.).

Global port agents (or port operators) provide the facilities as well as the strategic planning for infrastructure investment (Rodrigue et al. 2013). Their integration process involves mergers and acquisitions of existing terminals or the construction of new terminal facilities. Reported by the Lloyds list in 2018, the world's top five port operators, which handled more than 50% of the global trade flows (counted TEUs for the year 2018): (i) COSCO (Beijing, People's Republic of China); (ii) Hutchison Port Holdings (Hong Kong, People's Republic of China); (iii) APM Terminals (The Hague, Netherlands); (iv) PSA International (Singapore); (v) DP World (Dubai, United Arab Emirates) (Lloyd's List 2018) (Fig. 5.26).

DP World has therefore marked itself as a leading global terminal operator as trade enabler with an ambitious expanding policy, seeking to exploit shareholder value by enriching its portfolio with competing destinations and state of the art infrastructures and consolidate its position in the global supply chain network while increasing its economic growth and promoting environmental sustainability in the constantly. DP World has developed a prolific of 78 destinations—port terminals—in 40 countries in 7 world regions along with port terminals, logistical and industrial parks, free trade zones, maritime services, etc. (as of year 2017) 19 of DP World's terminals are ranked among the top 100 world busiest container ports. Figure 5.27 shows the worldwide location of this company's terminals throughput years: in the period between 1970s till mid-2000s (the white circles) DP World consolidates its footprint in its neighbouring region: Middle East and then Eastern Asia; it then reached out to Europe, Americas and Australia.

The worldwide geography of DP World's presence is illustrated in Fig. 5.28 to highlight the company's strategy in enriching its portfolio with more destinations in every region. Asia has the majority of ports across the rich portfolio given the

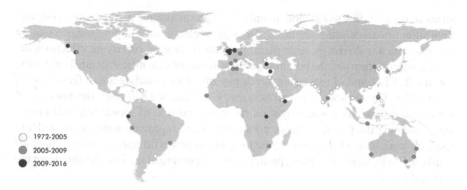

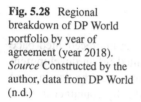

Fig. 5.27 Global expansion: DP World terminals by year of agreement (year 2017). *Source* Constructed by the author, data from DP World (n.d.)

Fig. 5.28 Regional breakdown of DP World portfolio by year of agreement (year 2018). *Source* Constructed by the author, data from DP World (n.d.)

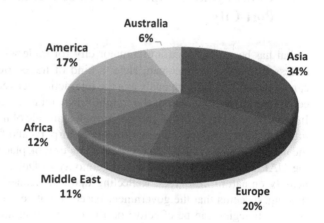

high volume of commodities produced in the region and then distributed around the globe. Europe follows with 20% having significant tradition in trade, logistics and ports function in many centuries. America, Africa, Middle East and Oceania are following with percentages 17, 12, 11 and 6% accordingly.

Studying the internationalization, geographical expansion and portfolio diversification strategies of DP World, as a leading global terminal operator, through acquisitions, mergers and reorganization of assets (Notteboom and Rodrigue 2012), gives rise to the question on the particular impact of such international operators on spatial patterns in terms of infrastructural, institutional and settlement changes. On this note, it is interesting to track, for this particular case of DP World, the eventual processes of *"Dubaization"* (i.e. the transfer of Dubai-like models of urbanization and regionalization attached to the development or expansion of port hinterland). Two recent academic research studies (see the two master theses co-supervised by the author of this book: Exarchou 2019; Mattana 2018) have made attempts to trace such impacts, testing four main hypotheses regarding DP world's presence and influence through series of case-studies worldwide, mainly through qualitative methods: (i) no significant effect or trace are recognized; (ii) minor effects on the port-regionalization process, which is driven by other plans, policies or factors; (iii) major effects on the urbanization pattern and the port hinterland in terms of new infrastructure besides the port itself, FTZs and of real estate developments in surrounding areas; (iv) eventual exports the 'Dubai model' and subsequent Dubaization. Although the findings of these studies have shed some light on the potential role of global terminal operators in shaping the port hinterland, urbanization and regionalization, more in-depth research work in needed to capture the tangible of powerful international companies in urban development processes, particularly concerning at the interface between the port and the city.

5.7   Conclusion: Competitiveness of Dubai Global Hub Port-City

Dubai has been developed into a major city for trade with an unprecedented pace in less than 40-year time span. Here, a kind of trade-oriented urban development can be recognized through analysing the interaction between some urbanization and economic factors with that of the freight flows, more specifically container traffics. Indeed, this oil-city has been shaped through a mosaic of megaproject trade-oriented infrastructural developments: maritime port, airports and free zones. Understanding the national context in which these local changes took place is also essential: in brief, the UAE owes greatly its diversified economy to Dubai, as the country yet relies heavily on natural resources. Reflecting upon the leader's vision towards hub-city making, it seems that the government-based initiatives coupled with internationalization strategies can be effective on the one hand in attracting FDI for the local growth, and on the other hand in consolidating its position as a regional and global transshipment and logistics hub within the global network. Henceforth, key drivers behind its growth and sustaining its competitiveness can be summarized as follows:

- Strategic location at the crossroad of the major shipping line and trading routes
- Growing, diversified and stable economy
- Growing urbanization (with a multinational demographic pattern)
- Provision of modern infrastructures (ports and airports)
- Growing international trade: increasing transshipment, added value industries, and establishments in free trade zones
- Multimodal logistics capabilities
- The active role of the government and varied public institutions.

References

Afrikian G (2017) Dubai multimodal transportation and logistics cluster platform. In: Presentation at the logistics spring conference, Luxembourg, May 2017

Akhavan M (2015) Port development and port-city interface dynamics. PhD dissertation. Politecnico di Milano

Akhavan M (2019) Gateway: revisiting Dubai as a port-city. In: Molotch H, Ponzini D (eds) The new Arab urban. Gulf cities of wealth, ambition and distress, pp 175–193. https://doi.org/10.18574/nyu/9781479880010.003.0008

Albawaba (2011) Department of economic development launches logistics cluster platform to drive growth of logistics. Available at https://www.albawaba.com/department-economic-development-launches-logistics-cluster-platform-drive-growth-logistics-376346

Altshuler A, Luberoff D (2003) Mega-projects: the changing politics of urban public investment. Brookings, Washington

Bagaeen S (2007) Brand Dubai: the instant city; or the instantly recognizable city. Int Plan Stud 12(2):173–197

Bottasso A, Conti M, Ferrari C, Merk O, Tei A (2013) The impact of port throughput on local employment: evidence from a panel of European regions. Transp Policy 27:32–38. https://doi.org/10.1016/j.tranpol.2012.12.001

Buckley M, Hanieh A (2014) Diversification by urbanization: tracing the property-finance nexus in Dubai and the Gulf. Int J Urban Reg Res 38(1):155–175. https://doi.org/10.1111/1468-2427.12084

Chen X (1994) The changing roles of free economic zones in development: a comparative analysis of capitalist and socialist cases in East Asia. Stud Comp Int Dev 29(3):3–25

Clark X, Dollar D, Micco A (2004) Port efficiency, maritime transport costs, and bilateral trade. J Dev Econ 75(2):417–450. https://doi.org/10.1016/j.jdeveco.2004.06.005

Cuthbert J (2011) The 40-year history of UAE logistics: part one—ports. Retrieved from https://www.arabiansupplychain.com/article-6889-the-40-year-history-of-uae-logistics-part-one--ports/1/print/

Davidson CM (2008) Dubai: the vulnerability of success. Columbia University Press, New York

Davis M (2006) Fear and money in Dubai. New Left Rev 41:46–68

DP World (n.d.) Available from https://web.dpworld.com/

DP World Annual Report (n.d.) Several years. https://web.dpworld.com

Dubai Chamber of Commerce and Industry (2004) Foreign trade of Dubai's free zones 2001–2002. Data Management & Business Research Department, Dubai

Dubai Customs. Available at https://www.dubaicustoms.gov.ae/en/Pages/default.aspx

Dubai Statistics Center. Available from https://www.dsc.gov.ae/en-us

Ducruet C (2009) Port regions and globalization. In: Notteboom TE, Ducruet C, de Langen P (eds) Ports in proximity: competition and coordination among adjacent seaports. Ashgate, Aldershot, pp 41–53

Ducruet C, Lee S-W, Ng AKY (2010) Centrality and vulnerability in liner shipping networks: revisiting the Northeast Asian port hierarchy. Marit Policy Manag 37(1):17–36. https://doi.org/10.1080/03088830903461175

El Hag S, El Shazly M (2012) Oil dependency, export diversification and economic growth in the Arab Gulf States. Eur J Soc Sci 29(3):9

Elsheshtawy Y (2010) Dubai: behind an urban spectacle. Routledge, London

Emirates Competitiveness Council (2012) Policy in action: Dubai trade—building competitive advantage through collaboration, Issue 03, Jan 2012. United Arab Emirates

Encyclopedia Britannica (n.d.) Free-trade zone. International trade. Available at https://www.britannica.com/topic/free-trade-zone

Exarchou E (2019) The impact of global terminal operators on port infrastructure and urban development. The case of Dubai ports world as a trade enabler. Postgraduate dissertation, Politecnico di Milano

Federal Competitiveness and Statistics Authority. Available at https://fcsa.gov.ae

Flyvbjerg B (2005) Design by deception—the politics of megaproject approval. Harv Des Mag 50–59

Foster H (2002) Design and crime and other diatribes. Verso, London

Government of Dubai (n.d.) Dubai plan 2021: Dubai strategic plan 2015. URL https://www.dubaip lan2021.ae/dsp-2015-2/. Accessed 10 Sept 2017

Grobar LM (2008) The economic status of areas surrounding major U.S. container ports: evidence and policy issues. Growth Change 39(3):497–516

Grubel H (1983) Free market zones-deregulating Canadian enterprise. Fraser Institute, Vancouver

Hertog S (2019) A quest for significance. In: Molotch H, Ponzini D (eds) The new Arab urban. Gulf cities of wealth, ambition and distress. New York University Press, New York, pp 175–193. https://doi.org/10.18574/nyu/9781479880010.003.0008

Hoyle BS (1972) The port function in the urban development of tropical Africa Colloques Internationaux du CNRS, La croissance urbaine en Afrique noire et a Madagascar, Tome II, Editions due CNRS, Paris, pp 705–718

Hoyle BS, Hilling D (1970) Seaports and development in tropical Africa. Macmillan, London

Hoyle BS, Hilling D (1984) Seaport systems and spatial change: Technology, industry, and development strategies, p 481. Wiley, New York

Hvidt M (2009) The Dubai model: an outline of key development-process elements in Dubai. Int J Middle East Stud 41(3):397–418. https://doi.org/10.1017/S0020743809091508

Hvidt M (2013) Economic diversification in GCC countries: past record and future trends. Kuwait programme on development, governance and globalisation in the Gulf States (27). London School of Economics and Political Science, London

International Monetary Fund—IMF. Available from https://www.imf.org/external/data.htm

Jacobs W, Hall PV (2007) What conditions supply chain strategies of ports? The case of Dubai. GeoJournal 68(4):327–342. https://doi.org/10.1007/s10708-007-9092-x

Jebel Ali Free Zone (2007) Association corporate brochure. Freedom Matters, Dubai

Jebel Ali Free Zone Authority (n.d.). Available at https://jafza.ae/

Johansson H, Nilsson L (1997) Export processing zones as catalysts. World Dev 25(12):2115–2128

Keshavarzian A (2010) Geopolitics and the genealogy of free trade zones in the Persian Gulf. Geopolitics 15:263–289. https://doi.org/10.1080/14650040903486926

Kumar BR (2013) The UAE's strategic trade partnership with Asia: a focus on Dubai. Middle East Institute. Available at https://www.mei.edu/publications/uaes-strategic-trade-partnership-asia-focus-dubai. Accessed 10 Oct 2014

Lee S-W, Song D-W, Ducruet C (2008) A tale of Asia's world ports: The spatial evolution in global hub port cities. Geoforum 39(1) 372–385. https://doi.org/10.1016/j.geoforum.2007.07.010

Lloyd's List (2018) One hundred ports maritime intelligence informa. Available online in www.llo ydslist.com/topports18

Mathur LK (1993) Reducing sourcing costs through foreign trade zones. J Gen Manag 19(1):64–75

Mattana LL (2018) The transnational patterns of development. Investigating the urban effects of Dubai ports world on port-hinterland and urban development. Postgraduate dissertation, Politecnico di Milano

Moulaert F, Swyngedouw E, Rodriguez A (2001) Large scale urban development projects and local governance: from democratic urban planning to besieged local governance. Geogr Z 89(2/3):71–84. https://doi.org/10.2307/27818900

Musso E, Benacchio M, Ferrari C (2000) Ports and employment in port cities. Int J Marit Econ 2(4):283–311

Nandkeolyar KH (2020) Jebel Ali Free Zone: everything you need to know about Jafza. Gulf News. Available at https://gulfnews.com/uae/jebel-ali-free-zone-everything-you-need-to-know-about-jafza-1.1577097555736. Accessed 12 Feb 2020

Notteboom T (2013) Cargo volumes in the European port system. Port Technology International—Edition 59. Available from https://www.porttechnology.org/images/uploads/technical_papers/Theo_Notteboom_PT59_LR.pdf

Notteboom T, Rodrigue J-P (2012) The corporate geography of global container terminal operators. Marit Policy Manag 39(3):249–279. https://doi.org/10.1080/03088839.2012.671970

Papadopoulos N (1985) The free trade zone as a strategic element in international business. Can Bus Rev 12(1):51–55

Papadopoulos N (1987) The role of free zones in international strategy. Eur Manag J 5(2):112–120. https://doi.org/10.1016/S0263-2373(87)80074-9

Ponzini D (2011) Large scale development projects and star architecture in the absence of democratic politics: the case of Abu Dhabi, UAE. Cities 28(3):251–259. https://doi.org/10.1016/j.cities.2011.02.002

Ramos SJ (2010) Dubai amplified: the engineering of a port geography. Ashgate

Rodrigue J-P, Comtois C, Slack B (2013) The geography of transport systems, 3rd edn. Routledge, Abingdon

Romero AT (1998) Export processing zones in Africa: implications for labour. Compet Change 2(4):391

Saadi D (2013a) Arabian Gulf states put oil money to good use with port expansions. Available from https://www.thenational.ae/business/shipping/arabian-gulf-states-put-oil-money-to-goood-use-with-port-expansions

Saadi D (2013b) UAE's Khalifa Port and Jebel Ali lead way in port developments. Available from https://www.thenational.ae/business/industry-insights/shipping/uaes-khalifa-port-and-jebel-ali-lead-way-in-port-developments

Sargent J, Matthews L (2001) Combining export processing zones and regional free trade agreements: lessons from the Mexican experience. World Dev 29(10):1739

Scott A (2014) Dubai 33-year growth rate was world's fastest. Available at https://www.thenational.ae/business/dubai-33-year-growth-rate-was-world-s-fastest-1.275416

Swyngedouw E, Moulaert F, Rodriguez A (2012) Neoliberal urbanization in Europe: large-scale urban development projects and the new urban policy. In: Spaces of neoliberalism: urban restructuring in North America and Western Europe, pp 194–229. https://doi.org/10.1002/9781444397499.ch9

Tansuhaj PS, Gentry JW (1987) Firm differences in perceptions of the facilitating role of foreign. J Int Bus Stud 18(1):19–33

Tansuhaj PS, Jackson GC (1989) Foreign trade zones: a comparative analysis of users and non-users. J Bus Logist 10(1):15–30

UAEstablishment (n.d.) Dubai free zones. Available at https://www.uaestablishment.com/en/dubai-free-zones/

UNCTAD Port Section (1996) UNCTAD monographs on port management. New York and Genoa

United Nations (2005) Free trade zone and port hinterland development. United Nations ESCAP

WAM Dubai (2010) Sheikh Ahmed bin Saeed Al Maktoum inaugurates Dubai logistics corridor. Available at https://wam.ae/en/details/1395228768688

World Bank (2012) Doing Business 2012: Doing Business in a more transparent World. World Bank, Washington, DC

World Bank Database, online. Available https://www.worldbank.org

World Shipping Council. Available from https://www.worldshipping.org/about-the-industry/global-trade/top-50-world-container-ports

WTO-Trade Policy Review (2012) Report by the secretariat United Arab Emirates. WT/TPR/S/262/Rev.1, 3 May 2012. Available at https://www.google.com/url?sa=t&rct=j&q=&esrc=s&source=web&cd=2&ved=2ahUKEwiw36ngxZPpAhVM_SoKHVtoC2YQFjABegQIBRAB&url=https%3A%2F%2Fwww.economy.gov.ae%2Fenglish%2FUAE-Office-WTO%2FUAE-and-WTO%2FDocuments%2FTRADE%2520POLICY%2520REVIEW%2520Report%2520by%2520Secretariat%25202012.pdf&usg=AOvVaw3SKvqj7LPcx-csgfAqYDkh

Xinhua (2017a) China remains top trade partner of Dubai-based economic zone. Available at https://www.xinhuanet.com//english/2017-08/13/c_136522808.htm

Xinhua (2017b) UAE's logistics sector expected to boom report. Available at https://www.xinhua
net.com/english/2017-08/24/c_136550058.htm
Ziadah R (2018) Transport infrastructure and logistics in the making of Dubai Inc. Int J Urban Reg
Res 42(2):182–197. https://doi.org/10.1111/1468-2427.12570

Chapter 6
Conclusions. Future of Port Geography in the Developing World

6.1 Critical Review of the Arguments

Many cities originally settled by a port have historically built a complex relationship with their port. As discussed in Chap. 1, the economic life of such cities was highly dependent to its port as a source of employment and trade with the international market. Nevertheless, in the recent decades, as a consequence of globalization and containerization, the dominant trend for Western countries has been the growing level of disconnection between the port and its city. The stark gap in the literature on port-city studies is identified: there is a bias in research studies on this topic that is mainly the Western World, with a recently growing literature on some East-Asian cases. Although some academics have emphasized the importance of recognizing economic growth in Asian countries because of the increased amount of freight movement from Europe-East trade (Musso and Parola 2007), but the shipment revolution and the increase in logistic centres in the Middle East still remain relatively unexplored. This book thus has made attempts to present different strand of literature on the spatial relation between the port and the city at different scales in a globalized world, as discussed in Chaps. 3 and 4. The aim is to contribute to the vast existing literature, by the means of case-study approach coupled with qualitative and qualitative analysis presented in Chaps. 4 and 5.

As reviewed in Chap. 2, many conventional and recent studies have used models providing general characteristics for explaining the evolutionary trend in port-city interface. Most spatial and geographical models for the Western developing countries conclude with the idea of weakening port city relation after undergoing similar phases of development. Asian-based models, in contrast, are characterized in the continuation of port activities in close proximate to the city core. Modernized shipping and logistics technologies nevertheless require new infrastructures and more land for hinterland expansion; this has resulted in common demand for land in all cases worldwide. Advancements in transportation technologies have affected modern changes in the global, regional, and local economic systems. Such changes include containerization, and increased freight flows, which require more studies connected to maritime trade and logistics (Hoyle and Knowles 1998; Taaffe et al. 1963; Vanelslander and Musso 2015). The circumstances of these aspects were discussed in Chap. 3 with reference to the literature on new transport geography of logistics and freight flows. Referring again to the results of O'Connor's study (2010), global cities with vast logistics activities are those that have both specialized seaports and airports. Here, Singapore and Dubai—as two global logistics centre—are ranked as top global cities, Alpha+ City ranking (GaWC n.d.).

On the basis of the conventional port-city development studies of the Western and Eastern world an evolutionary model was introduced, in Chap. 4, and applied to the case of Dubai, to explain its changing port and city relations in time and space. Table 6.1 summarizes the main feature of each phase, in terms of the main economic activities, key governmental strategies and policies, urban planning practices, main infrastructural characteristics, general urban development features and the port-city interface.

Dubai has, indeed, been successful in attracting expatriates: *phase one*, the merchant came and remained in the city to develop trade and commerce; *phase two*, low-skilled labours were attracted to respond to the demands brought by the construction boom; *phase three*, high-skilled professionals essential to develop the diversified economy. However, in the *phase four*, the current era, Dubai is still lacking professionals to meet the requirements of some essential sectors of the knowledge-based economy. This is one of the main challenges, shadowing Dubai's success in becoming a major 'logistics-hub port city'. Concluding Dubai's specific pattern of development, it is worth underlining the importance of immigrant labour force in determining the socio-economic pattern of the city. It has been accepted that the great influx of expatriates has been one of the main factors contributing to the urban development in this city (Pacione 2005). As an outcome of this massive flow of immigrants, it should be noted the significant demographic profile of Dubai. The share of non-nationals to the native population is the highest in the world: currently Emiratis account for less than 8% of the total population. The Asian and Arab expatriate dominant the city-state. The city also hosts a high number of people from the Western world, preliminary being the British.

In Chap. 5 the test bed success of Dubai development into a major transshipment hub and logistics centre was explained through government-led strategies in diversifying the economy. Data on trade, commodity flows (volume of container) was analysed with respect to some factors on urbanization (population) and economic growth (GDP growth and share of main activities in GDP). Urbanization in this case is preliminary generated and then elevated by means of trade-oriented infrastructures. To better comprehend this trajectory, Fig. 6.1 clearly shows this rapid development in a short period, highlighting the urbanized areas, maritime ports and airports, free zones and the main road networks, in three main stages: in the year 1975, with only one port and airport and the urban settlements are agglomerated around the Creek (the historic centre); in the year 1990, one can recognize the linear growth of the city towards the Jebel Ali area along the main road infrastructure and parallel to the coastline; the year 2015 one can acknowledge a city made of mosaics of megaprojects and free zones, which is characterized by specialized expansion of port-hinterland

Table 6.1 Main features of the four-phase Dubai port-city development pattern

Four-phase of Dubai port-city development	Period	Main economic activities	Key governmental strategies and policies	Urban planning practices	Main infrastructural characteristics	General urban development features	Port-city interface
Phase I: The fishing village and the advent of a free port	1900s–1950s	Fishing and pearl trade	Tribal governance; free trade policies	No formal planning strategy	Indistinct urban infrastructure; minor port (Creek) expansion	Minor settlement around the Creek	Two separate urban settlements around the Creek
Phase II: The entrepôt port-city	1960s–1970s	Natural resources (oil and gas)	Centralized decision making; investment into trade-based infrastructure	First and second master plan; two major modern port planning	Creek expansion; road system; construction of two main two ports; docks	Moderate development along the Creek	Expansion around the Creek and the new port. city centre becomes more congested
Phase III: The transshipment hub port city	1980s–1990s	Trade and real estate; port-related industries	Privatization and liberalization of the governance; mega project development strategy	Comprehensive development planning; structural planning	Free trade zone development—Dubai-Abu Dhabi (Sheikh Zayed Road)	Massive real estate development along the coastline and growth corridor of Dubai-Abu Dhabi axis	Two different port-city interface s at either end of the city: the historic centre and the peripheral centre

(continued)

Table 6.1 (continued)

Four-phase of Dubai port-city development	Period	Main economic activities	Key governmental strategies and policies	Urban planning practices	Main infrastructural characteristics	General urban development features	Port-city interface
Phase IV: The logistics hub port city	2000s–present	Finance, logistics, tourism, etc.	Integrative development strategy; public–private partnership	Strategic planning	Multi model trade, transport and logistics platform	Mega project development continues, with a focus on tourism	The historic interface (Port Rashid) is dedicated to urban activities (residential, business and tourism). Jebel Ali interface is specialized in logistics activities

Source Author

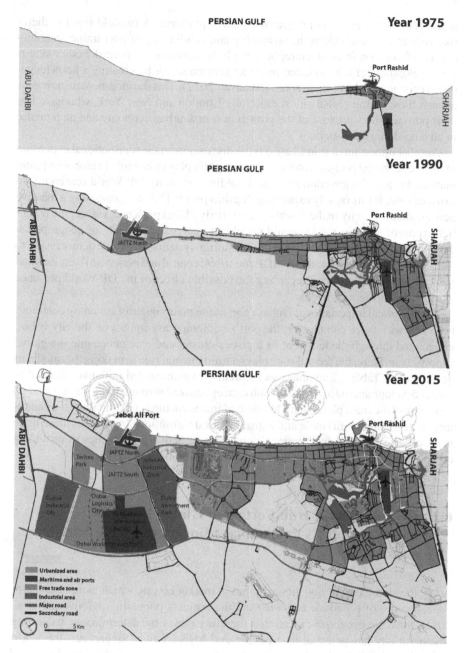

Fig. 6.1 Dubai urban development and expansion of trade-oriented infrastructures (main free zones, maritime ports and airports) for three years of 1975, 1990 and 2015. *Source* Author

at the Jebel Ali area as an integrated logistics platform. A twofold trend is therefore recognized: one side is the sustaining and reinforcing of port traffic, enabling the transformation from an entrepôt into a hub transhipment port; the other side is the development of the advanced producer service sector by creating a knowledge-based economy and free trade zones (Akhavan 2017). This development pattern goes against those of top global cities, especially London and New York, who have lost their port-activities because of the growth of a new urban economy and an increase in advanced producer services.

Another important note in Chap. 5 is the discussion on the transnational expansion of the Dubai-based port-operators, which currently plays a key role in global shipping market. In fact, the government-owned holding company DP World represents an extreme case, for its rapidly expanding flagship port in Dubai—generating a massive new quadrant of a city in the desert—and its truly global presence. Hence, the potential export of a Dubai model would possibly target a large variety of geographical destinations (different institutional and planning systems, from full democracies to dictatorships; high-GDP and low-GDP countries; central and peripheral locations and markets, etc.). The question regarding the possible effect of the DP World presence worldwide, still remain opens.

The historical importance of Dubai's port as the main source of economy continues till today with more complexity; the port's contemporary impact on the city is best understood through the lenses of its logistics cluster and emergence into the global supply chain. Here, the federal state plays a fundamental role in making this model of development. Table 6.2 summarizes the main government-led initiatives that led to Dubai development into a logistics hub conceptualized through three main gears: (i) provision of the main physical trade-oriented infrastructures, (ii) trade/logistics ancillary infrastructures, (iii) trade and logistics-based institutions. As well as these three main pillars, the local (and national) government actively intervenes in providing complementary policies (trade, tax, business environment, etc.).

6.2 Development of Hub-Ports into Global Logistics Centres: Dubai's Potential as 'Singapore' of the Middle East?

A key location factor of logistics activities is market access, which includes a major concentration of economic activities and the region's population (Gouvernal et al. 2011). Strategic geographic location is therefore one of the determinants in defining whether a city has potential as a trading and logistics hub. Singapore—the world leader logistics hub—is within the South Asian tigers, controlling the Straits of Malacca, where a large amount of world trade passes. Dubai is also strategically located, within the rich Persian Gulf States and close to the Strait of Hormuz, on the new Southern Silk Road between Asia, Africa, and Europe; the city-state has been a gateway to the international market of 105 billion in the neighbouring regions

Table 6.2 Dubai key government led initiatives to develop a logistics hub

Stage	Year	Key initiatives
Main trade-based infrastructures	1900s	The Creek a free port
	1959	Dubai International Airport
	1961	Dredging Dubai Creek
	1972	Port Rashid opened
	1978	Port Rashid expansion
	1979	Jebel Ali Port opened
	2010	Al Maktoum International Airport opened for cargo flights
	2014	Jebel Ali Port inaugurated a new terminal (T3)
Ancillary trade- and logistics-based infrastructures	1983	Dubai Dry Dock
	1985	Jebel Ali Free Trade Zone opened
	1990	Jebel Ali Free Trade Zone expanded
	2007	Stage one Jebel Ali Port expansion finished
	2008	Dubai Maritime City inaugurated
	2009	Dubai Industrial City inaugurated
	2010	Dubai Logistics Corridor launched
Trade- and logistics-based institutions	1957	Dubai Municipality established
	1985	Dubai Sky Cargo
	1991	Dubai Port Authority (DPA) founded
	1999	Dubai Port International (DPI) founded
	2001	Ports, Customs and Free Zone Corporation (PCFZ) launched
	2005	DP World was formed (DPI merged with DPA)

Source Gathered by the author

(Persian Gulf, Middle East/Eastern Mediterranean, Central Asia, Africa and Asian sub-continent) since the 1980s.

In the literature, Singapore is considered an exceptional transhipment port that has been successful in attracting logistic activities and developing into a global logistics centre (Slack and Gouvernal 2016). Its success could only occur because of a combination of rarely replicable conditions, such as strategic geographic location, transhipment traffic size, scale of hinterland traffic, government policies for trade facilitation measures, and the high quality of both labour and IT in the city-state. Any of these factors, individually, would not be able to generate significant logistics at any intermediate transhipment facility, but together they helped Singapore become the world's leading global logistics centre.

A recent study has conducted comparative case-study study approach to evaluate Dubai's competitiveness as a logistics hub compared to Singapore, based on the following indicators: location; port container and air cargo handling; basic socio-economic indictors (population, GDP, GDP per capita and share of main sectors in the economy); infrastructures and transport facilities; ease of doing business; and more importantly the Logistics Performance Index[1] (LPI) for the period 2007–2016 (Akhavan 2017). Both Singapore and Dubai are able to manage a relatively high amount of cargo flow by hosting major container ports and airports in the region, even if the population density and GDP in Singapore is much higher than that of Dubai. Significantly less data on UAE (Dubai) exists (compared to data on Singapore), but both the trade-infrastructural investments and increasing freight traffic (both sea and air) and also the LPI in the recent decade prove that Dubai's ambition is transforming the city from a regional trading hub into a major global trading hub.

Based on the most recent report from the World Bank's LPI (Arvis et al. 2016), many oil exporting countries, such as Qatar (30), Bahrain (44), and Saudi Arabia (52) seem to perform poorly in terms of logistics. One reason for this poor performance may be because of the high share of oil export, and therefore the lack of initiatives and pressure to transport in other trade sectors. Conversely, the UAE has been able to perform strongly in the logistics sector with its diversified non-oil export strategies. The LPI index proves that UAE is increasing its competitiveness in an improvement trend that surpasses that of Singapore. But the overall performance over the years has shown that there still exist large gaps between the two country cases; most of these gaps emerge because of the difference in custom, logistics quality, and ease of tracking. Regardless of these gaps, the UAE (with Dubai contributing most) has emerged in the top 15 countries worldwide in the logistics cluster, maintaining the best logistics and trade facilitation performance. This allows the UAE to outperform many emerging economies like China, India, Brazil, and the Russian Federation.

Both Singapore and Dubai are examples of how cities have optimized port land-use, with regard to planning spaces within the port for future use in key sectors. For example, shipyard activities and ship repair facilities create economies of scale and therefore contribute to the growth of the maritime system within the port (Merk 2010). The maritime clusters in both Singapore and Dubai have developed activities that complement the activities and policies that connect the port with the immediate hinterland. Consider the project of the Dubai Logistics Platform. Singapore is one of the best examples of how proactive policies can attract high value-added companies to the region and help to create an international maritime centre. Government policies, like customs, tax benefits and tariffs, are of great importance when studying countries and their logistics and trading performances. While Singapore has not implemented tax-free-policies like Dubai, it is one of the leading nations with regard to the openness of their economy. The World Bank measurement 'Ease of Doing Business' provides a wholistic picture of the regulatory environment created for both the opening and operation of a local firm. Singapore has headed the 'Ease of Doing Business' ranking

---

[1]The World Bank's Logistics Performance Index (LPI) is the primary tool for the in-depth investigation of logistics performance for countries and is therefore able to benchmark results against 150 countries. LPI is complementary to international competitive indicators, like the World Bank's Doing Business measures and the World Economic Forum's Global Competitiveness Index (Arvis et al. 2007).

within the past decade, ranked 1st preferred destination for doing business. Dubai (UAE) was ranked 22nd in 2014; yet has significantly improved from its position as 51st in 2006 (World Bank Group 2014).

According to Majdalani et al. (2007) Dubai is best positioned to become the next global logistics hub if it can create the following elements: (i) an economic environment to attract foreign firms and investors; (ii) a large free trade zone (JAFZA) adjacent to the maritime-port and airport; (iii) highly competitive handling charges; (iv) a large expatriate population. Nevertheless, other studies on Dubai's potential as a logistics hub reveal that Dubai is behind the top ranked logistics performing countries, which needs more investment and development (Fernandes and Rodrigues 2009). Dubai's shortage is also connected to the lack of human capital. Whether Dubai can gain a strong position to become the 'Singapore of the Middle East' is yet to be answered, as other competitors continue to develop throughout the region—such as the ports of Qatar and Bahrain—and the city's full potential has yet to be reached (Akhavan 2017).

6.3 Closing Remarks and Future Research

In addition to the discussion in the previous section, for emerging hub port-cities, like Dubai, the Singapore experience is the best example and benchmark to understand the gaps they face and needs that have yet to be met to make their hubs more successful. Dubai has gained success in following a development path similar to Singapore. The hinterland development in the Dubai region has also contributed to the port-city's success. The recent projects to create a multimodal logistics platform next to the main maritime port is a game-changing achievement, which has helped to establish the UAE as the best performing service-provider in the region. Dubai is supported by leading maritime infrastructure and airports but falls behind in the development of a well-linked hinterland through a land transport system and logistics platform in order to form a more efficient system for managing the supply chain.

The topic of rapid urbanization in the globalized world has been studied extensively, and while each case is different, there are common trajectories among a set of cities especially sharing socio-economic and geographical features. Nevertheless, Dubai's rapid urban development is in many ways unique, as the city demonstrates some particular attributes that makes it stand out among other emerging cities of the Middle East. Another important factor that distinguishes Dubai, especially from the similar post-oil cities of the region, is its hub-strategy and intermediary role in the global economy. Although the city has been investing on developing the manufacturing and industrial sector, it relatively produces little and instead focuses on mediating flows of people, goods, capital, knowledge and ideas. Hence Dubai becomes an intermediately node within the complex network of the regional and global connections. Following the statement by Castells (1989: 142): '*Flows, rather than organizations, become the units of work, decision, and output accounting*', the spaces of flows are the key factor in defining Dubai's pattern of rapid urban growth in a global economic environment.

From an urban planning perspective, it is expected that the research findings will provide a comprehensive basis for private and public actors dealing with the port and city development issues. Having knowledge about the role of global port operators in several ports worldwide and their hinterland development is essential for port-authorities (public and private actors) and local government (public actors) in designing and applying effective projects and policies. Indeed, Dubai has been successful in mediating the regional flows, as a regional trading/transportation hub, and becoming a major player in global supply chain, yet there is a need to reorganize and implement an integrated strategic planning that benefits and supports the human capital and the socio-economic status of the city. Henceforth, this study calls for future research to understand to what extent Dubai is lacking human capital on the one hand, and more importantly how the incoherent planning of port development and the surrounding urban areas has affected the socio-economic condition of the city and the quality of life of the residents.

References

Akhavan M (2017) Evolution of hub port-cities into global logistics centres: lessons from the two cases of Dubai and Singapore. Int J Transp Econ 44(1):25–47. https://doi.org/10.19272/201706 701002

Arvis J-F, Mustra MA, Panzer J (2007) Connecting to compete : Trade logistics in the global economy. The logistics performance index and its indicators. The International Bank for Reconstruction and Development/The World Bank, Washington, Charlier

Arvis J-F, Mustra MA, Ojala L, Shepherd B, Saslavsky D (2016) Connecting to compete-trade logistics in the global economy: the logistics performance index and its indicators. World Bank, Washington

Castells M (1989) The informational city: information technology, economic restructuring, and the urban-regional process. Basil Blackwell, Oxford

Fernandes C, Rodrigues G (2009) Dubai's potential as an integrated logistics hub. J Appl Bus Res (JABR) 25(3):77–92

GaWC (n.d.) Globalization and World Cities research group. Available at https://www.lboro.ac.uk/gawc/

Gouvernal E, Lavaud-Letilleul V, Slack B (2011) Transport and logistics hubs: separating fact from fiction. In: Hall P, McCalla R, Comtois C, Slack B (eds) Integrating seaports in trade corridors. Ashgate, Farnham, pp 65–79

Hoyle BS, Knowles RD (1998) Modern transport geography. Wiley, Chichester

Majdalani F, Koegler U, Kuge S (2007) Middle East transport and logistics at the crossroads

Merk O (2010) The competitiveness of global port-cities: synthesis report. OECD port-cities programme. Available at https://www.google.com/url?sa=t&rct=j&q=&esrc=s&source=web&cd=1&ved=2ahUKEwjSy4nPqanpAhVus4sKHX1PDlIQFjAAegQIBBAB&url=https%3A%2F%2Fwww.oecd.org%2Fcfe%2Fregional-policy%2FCompetitiveness-of-Global-Port-Cities-Synthesis-Report.pdf&usg=AOvVaw3pkUIrwCg2x6zPXyh4MK3j

Musso E, Parola F (2007) Mediterranean ports in the global network: how to make the hub and spoke paradigm sustainable? In: Wang J, Olivier D, Notteboom T, Slack B (eds) Ports, cities, and global supply chains. Ashgate, Aldershot, pp 89–104

O'Connor K (2010). Global city regions and the location of logistics activity. J Transp Geogr 18(3):354–362. https://doi.org/10.1016/j.jtrangeo.2009.06.015

Pacione M (2005) City profile: Dubai. Cities 22:255–265. https://doi.org/10.1016/j.cities.2005.02.001

Slack B, Gouvernal E (2016) Container transshipment and logistics in the context of urban economic development. Growth Change 47(3):406–415. https://doi.org/10.1111/grow.12132

Taaffe EJ, Morrill RL, Gould PR (1963) Transport expansion in underdeveloped countries: a comparative analysis. Geogr Rev 53(4):503–529

Vanelslander T, Musso E (2015) Transport, logistics and the supply chain: how changes reshape the research agenda. Int J Transp Econ IJTE 42(1):11–15

World Bank Group (2014) Doing business. Available from https://www.doingbusiness.org/rankings. Accessed 20 Sept 2015

Printed in the United States
By Bookmasters